Vietnam

Without A Dream

by

Jerry L. Harlowe

NO MATTER WHAT YOU THOUGHT IT WAS, IT WASN'T
NO MATTER WHAT YOU THOUGHT IT WASN'T, IT COULD HAVE BEEN ~

BUT, WHATEVER IT WAS, OR WASN'T, IT REMAINS TOO PERSONAL TO REALLY DISCUSS... IT WAS VIETNAM, OR WAS IT?

SO, IT WAS 52 YEARS AGO THIS JULY 4th THAT I ARRIVED BACK AT 1425 HOPEWELL AVENUE, WHERE MY PARENTS LIVED. I WALKED IN, AND IT WASN'T... I SHOWERED, CHANGED AND GOT INTO MY '67 CAMERO SITTING OUT BACK... IT WASN'T, BUT I DROVE OFF HOPING I COULD FIND WHAT WAS... AND IN SOME WAYS, I'M STILL DRIVING

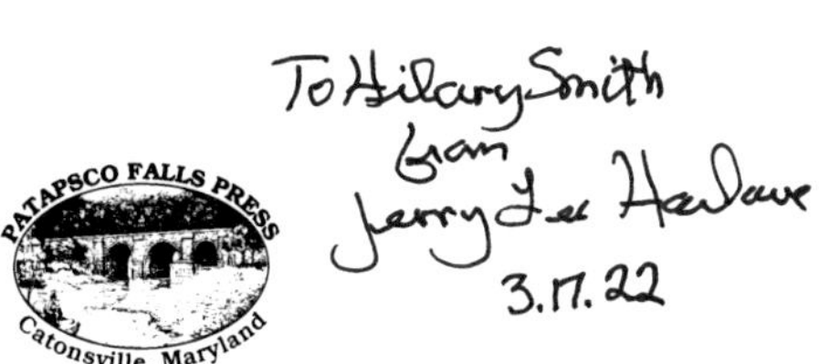

is published through cooperation of

and

Merrifield Graphics & Publishing Service
Book Design and Production
Baltimore, Maryland

ISBN 978-0-9634687-3-4

Contents

Prologue

I never started out to write a book, and I feel I have accomplished that goal with this loose collection of short stories and even shorter recollections. These are my tales and memories of Vietnam and later events subsequent to the war years that somehow always pulled me back to Vietnam.

My tour of duty started with my arrival at Pleiku Air Base in November of 1968 and, for most people, I suppose my tour of duty would seem to have ended in July 1970 when I left Pleiku for the last time during the war. My separation from the Air Force followed a few days later in the State of Washington and life began anew... or so I thought.

In these little stories I will introduce you to the Bird Colonel of Pleiku, Staff Sergeant Stokes and Martha Raye. I also want you to meet Kim of the Han Long Hotel, the model with the Hong Kong smile, Jenny of the California Bar, Vicki, Jane Fonda and maybe the dead and pickled Ho Chi Minh.

These are memories, my memories and my little stories.

In his brilliant book, *The Things They Carried,* author Tim O'Brien wrote of memories and what they represent. I've never read better:

> Forty-three years old, and the war occurred half a life-time ago, and yet the remembering makes it now. And sometimes remembering will lead to a story, which makes it forever. That's what stories are for. Stories are for joining the past to the future. Stories are for those late hours in the night when you can't remember how you got from where you were to where you are. Stories are for eternity, when memory is erased, when there is nothing to remember except the story.

O'Brien, Tim, *The Things They Carried,* Houghton Mifflin, Boston. 1990. p.38

Jerry Harlowe and W. Rogers, Pleiku AB, 1969

Vietnam.

Every sense rushed, shocked, drawn and quartered
till permeation was complete
and
reassuring
and
you sucked in the vivid color and cycle smoke,
green-brown mountains
and
too-tuned children
and
stepped over the beggars lining the confused streets
and
you were invigorated by it, sought it,
because its bedlam was sure and never confusing

Mara 1967 – 1969

Vietnam
Without A Dream

Jerry Harlowe, Pleiku Air Base, 1969

To my buddy, Denny, USMC, who quietly held me
as I cried over the death of my friend, Ron.

Pleiku Vietnam, 1968-1970

CHAPTER ONE

Part I: Welcome to Vietnam, Troop

How best to be welcomed to a "war-zone than with a good dose of contemptibly petty, insignificant nonsense... a.k.a., chicken-shit!

Also, there will be other short tales in here, as noted under Chapter One heading, as well as some comments on official breaks away from Pleiku: The infamous "R&R" trips, in and out of the country of Vietnam.

The moniker, "R&R" was short for Rest and Recuperation. It was an official time away from duty to allow the person to REST from the day to day stresses, and to RECUPERATE in the mind and body so that upon return, you would be ready to go at "it" again.

Most of the married guys would fly off to Hawaii to meet their spouses, while the other fellows would head off to Japan, Hong Kong, Manila, Taiwan, Singapore, Kuala Lumpur, Australia and a place or two that I have forgotten by now. The R&R would be remembered as "I & I" by most of the guys... that is, Intoxication & Intercourse. Need I say more?

I will also include a few thoughts on my experiences upon coming home in July 1970 after 21 months at Pleiku Air Base and vicinity. It was not a parade, just weird.

Of course, there are many things to write about, but these are highlights, flashes of time now gone, and these memories are fading with age.

Welcome to Vietnam

For those of you who are familiar with the military, or any large corporate structure, I am sure you have your own chicken-shit stories so indulge me as I start with the retelling of my favorite.

On my first full day at Pleiku Air Base, Thanksgiving 1968, I was walking up the hill of the main base road near the intersection with the chapel. As I approached the intersection, coming down the hill towards me was a dark blue Ford automobile. I instantly thought it was an Air Police vehicle, the same as I knew from Langley Air Force Base in Virginia, my previous duty assignment. All too late I spotted the wee plate on the grill of the car with a full colonel's bird on it. The car passed before I could salute... "SHIT!"

I kept walking in the hope that my infraction of military courtesy would be dismissed by the lifer behind the wheel. But within a fraction of that very thought, I heard the car stop behind me and recognized the tight whining of a Ford in reverse. I continued to walk on in what had suddenly become a fruitless hope of escape. The Bird-Man himself stopped next to me, leaned across the blue vinyl bench seat as he slammed the Ford into "park" and attacked me with a full automatic burst of ass-chewing for my trespass of military decorum.

I stood there as dumb as a water buffalo's turd on the side of the road. But when the colonel took a breath to reload his brain and recharge his mouth, I got a chance to surrender with my most perfect salute. I then proceeded to tongue-stumble out of my mental stupor with an explanation as to the failure of my observation and my subsequent neglect to salute. He listened.

I suspect the colonel's adjudication of my proffered defense was most likely tempered by the fact that I was sharp in my obvious stateside starched fatigues which surely accentuated the Gomer smile on my face. For sure, I was a certified Vietnam cherry boy.

The ass-chewing assault ceased as suddenly as it begun, and the colonel reminded me of the need to maintain military courtesy even in a war zone. He then clunked the Ford's transmission into drive, as I stood in a salute, and the base commander, the colonel, said: "Welcome to Vietnam."

Radio Peking

The first broadcast I ever heard out of Peking (Beijing), People's Republic of China, was in early 1969, very late in the evening. My fingers were slowly rotating the tuning knob of my portable radio, skipping all over the AM dial, catching stations here and there. I wanted to listen to anything except Armed Forces Vietnam radio, AFVN, as I just wasn't in the mood; maybe it was a letter from a stateside girlfriend or a case of the chow hall squirts, I don't remember. I wanted to listen to music that was specific to Vietnam even though I had no clue about the language. Stopping here and there along the dial, the music sounded like caterwauling and I found it rather irritating as well as boring. At age twenty-one I was convinced there was nothing the Vietnamese had to offer that could compare with Hendrix, The Doors, Joplin, Cream, and a host of other American musical demigods.

All was quiet that night. Clear sky, somewhat cool in the Central Highlands, but somewhere out there in the distance, out under the breathlessly beautiful black heavens of a Vietnam sky, people were getting pounded to bloody shit piles. I could hear it plainly, out there. Even over the caterwauling, I could hear it, but I did not know how to pray for them. I returned to twisting the radio knob... searching.

One station blasted in clear and caught my attention. It was clear, very clear. Someone had a huge, powerful transmitter somewhere and I was the recipient of its waves. In a very clear, hard and brash voice, the English voice, steeped in the taunting accent of Commie-Chinese, the announcer was proudly screaming that the next selection for this night would be: DOWN WITH AMERICAN IMPERIALISM! DOWN WITH RUSSIAN REVISIONISM! The announcer continued his ranting, I twisted the knob just slightly this and that way to secure the strongest signal, and I was informed that the next selection would be sung by 10,000 "People's Voices."

I was a little dumbfounded listening to the announcer and the blasting song. I had never been so eerily entertained by such masses of singing voices, nor had I ever been made aware of the legions of people who wanted to kill me just because I was an American. These guys on the radio really made the point.

Downtown Pleiku, Vietnam

As a kid I had seen all of the glory and gory World War II movies. In the racist writing of the time, there was always fanatical "Japs," "slopes," and a variety of "chinks" who were doing their damned best to kill Americans. I had never thought of myself as a Yankee Imperialist Pig worthy only of a grenade up my rear, a bayonet in the gut, or some other gory end of life. I figured there are two sides to a story, but that was not the point of those movies which were there to wave the flag and mow down the slant-eyed bastards! So, what was it that now made it MY TURN to face the "yellow hoard" without my permission or understanding?

I turned off Radio Peking and I listened once more to the far off explosions. It was the very first time I really tried to understand the sounds of war rolling through the breathlessly beautiful night in Vietnam. There was no answer: I have never understood.

Time Travel

In a collection of little things that I brought home from Vietnam was a weapon's description for the Soviet-made 107mm and 122 mm rockets. I had picked up the card in a small museum that the Explosive Ordnance Disposal (EOD) guys had assembled at Pleiku Air Base.

The card had its own form number, MACV-8800/1 (7-68) and consisted of several photos and weapon descriptions of those two rockets. The description of the 122mm rocket (the rocket of choice for use against Pleiku Air Base), was as follows:

Weight: w/fuse	101.85 lbs.
Length: w/fuse	75.4 in
Diameter:	4.8 in
Stabilization:	Fin and Spin
Color: warhead	Dark Gray
Motor:	Aluminum
Range: (maximum)	11,000 meters

During 1969 we were hit several times with these rockets and during one of these attacks I was driving a deuce-and-a-half, which I found it to be imperative to abandon just after the first hit... KRUMMPH! I killed the engine and quickly scanned for a ditch or bunker to get my butt safely in and lying low. I had the door open and just as I was starting to jump another rocket hit very near-by **KRUMMPH!**

I do not remember if I was on the truck step or flying through the air with the greatest of ease, but the explosion was the most amazingly beautiful thing I ever saw, and it was all in ultra slow motion. I witnessed the explosion as it burst and expanded out from the point of detonation, throwing ten thousand points of white-hot stars skyward. My mind was transfixed by the lethal beauty and time had stopped... or so it seemed.

122mm rocket body

The experience of the explosion was the longest micro fraction of a second that I have ever experienced. It all came to an end when I hit the ground and the slow concussion and sound waves caught up with me and slapped me back to real time.

There's not much more to say about this particular night and the explosion of that particular Russian rocket. It was launched by unseen people miles away who were hoping to destroy something or kill someone. With the passage of time I can now flippantly say, MISSED – ME! but, all in all, no one was killed that night at Pleiku Air Base and the damage was not high. As for the people who sent the rockets our way, they most likely had to contend with the gunship SPOOKY. The old aircraft was armed with three mini-guns that together could deliver 18,000 rounds a minute. The airplane was always in orbit around Pleiku at night and if a rocket launch flash was seen SPOOKY was there with pay-backs.

For me, I wonder sometimes about not getting hurt that evening and the fraction of a moment, of a second, that I saw the lethal beauty of the explosion... but all I need to do is close my eyes and think about it, and I'm there again in that frozen time travel.

Fish-Flopping

Back in 1966 I watched a guy in boot camp have a nervous breakdown while lying on his rack after a particularly stressful day. The fine details of the day and of the overall event have slipped past my grasp to fully recall, but I do remember it wasn't pretty. The medics were in attendance within minutes and I never saw the guy again. Just a couple of years later I watched another guy suffer a nervous breakdown and that one I remember well.

The young sergeant had just returned that day from Cam Ranh Bay. He had been shipped out to the large hospital there just a week earlier, suffering from a very bad case of nerves. The shrinks had either pronounced the sergeant fit to return to duty or considered him a shirker and sent him back to Pleiku with a bottle of soul-numbing drugs. Later that day, in the late evening, the sergeant's nerves "went to hell" with the first rocket's burst. The airbase suddenly had suddenly come under attack and was receiving a fairly high dose of 122's, Russian-made rockets, and for this nervous man it was beyond the burden he could bear. It was over for him.

High atop selected telephone poles, sirens endlessly wailed while the rockets continued to hit along the flight line and in the barracks area. In the middle of all this noise there was always a voice screaming over the PA system that Pleiku Air Base was under a rocket attack. No shit! Everyone in our barracks was running to the bunker just outside the barracks. The man always first to the door near my end of the barracks, Lehockey, was our resident frozen-screamer. It was his habit to run to the door and freeze. There he remained, arms spread, clinging to the doorframe, unintelligibly screaming. His frozen muscles would not move until the next guy behind him slammed into his back and got him started to the bunker, still screaming as he ran behind the blast wall.

I was far from being the first one down the steps into the dank interior of the thirty foot- long bunker, but I certainly was not the last either. However, by the time I jumped in the party had already started. Joints were softly glowing, bottles of Jack were being sucked on and passed around, and loud dirty jokes were coming fast and furious. Not too far from me I remember some lovesick soul who brought along his letter tape so as not to interrupt his listening to the mellow tones of his wife back in the world. Out of necessity he was playing it loud, oblivious to the rest of us but determined not to have her words drowned out by bursting rockets or loud laughing sons-of-bitches and potheads. His wife's voice was relating items of simple sanity – of his son's cold and that somehow she had gotten behind in the bills. She needed more money, could he send it? Of course she loved him and could not wait to meet him in Hawaii, but she needed the money now. He stroked the tape recorder, his wife at that moment, and he was disdainful of the human distraction at his feet. On the dank dirt floor the sergeant was fish-flopping and foaming at the mouth.

122mm rocket hit

My attention narrowed to the sergeant's face, contorted with a pain I couldn't understand. His rolling eyes were wide with wildness and a sinister red glow. Although I knew the glow was but a reflection of the overhead red light bulbs, I was transfixed by the show. It was gruesome and painful, but God only knows I had to watch. I remember being relieved that there were a couple of guys immediately attending his writhing, for I did not know the first thing to do to relieve this man's pain. I was just as helpless as he was at that moment. The man with the tape recorder stepped over the writhing, foaming body to remove himself from the distraction. He did not want to interrupt the climax of the tape. A few months later his wife dumped him.

After the "all clear" was sounded, the shaking, jellied-legged sergeant was half-carried, half-dragged away. I never saw him again. His locker was emptied and his bunk was stripped within a day. I suppose he had his limits, as we all must, and on that particular night he was simply pushed over the edge. He could stand it no longer and his overtaxed mind and body reacted. I didn't understand then and I don't understand today, but I still feel sorry for us both, for his obvious pain and for my inability to help.

I'm A Fan for Life

Although I did not spend many hours in the Pleiku NCO club, it is from being in that place that I have the most prized slip of paper in my collection of Vietnam memorabilia... "stuff" that I brought back with me. On the small folded sheet of paper is her simple autograph: "Bless You, Martha Raye." There was no official notice of the star's arrival and there was no "show" that she was about to perform for the troops. She had simply appeared as though from a cloud of the past.

I was writing a letter home from my table in the barracks when my buddy, Ernie, who had just returned from the NCO club, told me that Martha Raye was at the club. I asked him if she was planning on leaving soon (as though he should know her schedule) and got the appropriate " I dunno" reply I deserved. Without further delay I was out the door and short-cutting my way through the barracks area, speeding up the hill to what I called "The Juicer Club."

I entered the club and went for the large knot of men, knowing that is where Miss Raye would be... the center of attention, of course. Martha Raye was quite small. Short... very short and almost tiny. She was decked out in Special Forces camo, jungle boots and her green beret. I wonder quietly... where the hell they find clothes and boots that small?

Martha's face was movie star bright and the lady sported the largest mouth I'd ever seen. I guess my silly smile caused her to look at me briefly, she spoke kind nothings I don't remember, but I fell in love right then and there. I was a twenty-one year old troop who was "ga-ga" for a lady over twice my years and ancient by my measure of time. But she was one of those people who have that mysterious power, the power of just BEING that I cannot explain but when confronted by such as person, it is instantaneously and universally understood.

I tried speaking directly with her when her kindly gaze stopped and rested upon my wide-opened eyes. But, I stuttered like I was once again that six year old kid in speech class... damn! Somehow, in that blasted darn NCO club, smoke, beer and a band on stage pounding out some crazy song, in the midst of all of these rather large Green Beret Guys standing guard over Miss Raye, I

made myself understood to her that I wanted her autograph. She kindly complied with my wish, spoke some more words to me that I don't remember, and sent me on my way so she could attend to the other Gomers who were surrounding her. It's almost like time standing still.

As I write this quick memory, all these years later, I fully realize that at that special time in 1969, in the NCO club, in Vietnam, I had a direct connection with all those men who fell in love with Martha Raye in World War II and Korea. She was selfless and could, and did, knock the socks off of several generations of GI's. I count myself damn lucky to be her fan. What a lady.

Being There

I imagine there will always be debates about the "good, the bad, and the ugly" of the will and ability of the South Vietnamese Armed Forces to pursue a war against the Viet Cong and the North Vietnamese Army (NVA). Reams of materials have been written, pro and con, on the subject and much of it makes for lively debate among historians and veterans alike. And there are excellent empirical data, as well as vast amounts of anecdotal stories, available to support one's interpretations, decisions or denials. However, I will leave the academic arguments to those with more passion and hours of scholarship than I, but I will tender my own following anecdotal observation and throw it into the mix.

First, as a general observation, most permanent U.S. installations in Vietnam had the local population heavily involved in the day-to-day running of the bases. Among those assisting in the operation were the ubiquitous mama-sans. For the most part these Vietnamese women were honest and hard working, earning a very good wage, which had been presented to them by the available infrastructure of the American war machine. There are stories of Vietnamese nationals who worked both sides of the combatant fence. American employee during the day, and Viet Cong employee by evening. I'm not one to dispute the truth of that accusation but I wouldn't presume it of any of the mama-sans I knew. But the problem I wanted to bring out, if it's not already out there, is the corruption of the economy that we encouraged and actively engaged in with our Vietnamese employees. In our barracks each mama-san was responsible for, I believe, six GIs. Each of us were paying her $20 a month for the duties she performed such as laundry, cleaning, making up the racks, and endless other little things. Although $180 a month doesn't appear to be that much in today's economy, it was a fortune in 1969's Vietnam and was over half of what I cleared at the time as an E-4.

The problem, as I see it, is that we benignly participated in the corruption of the very Vietnamese mind, spirit and body politic by being so free and gracious with our cash. We were paying a mama-san a monthly salary of what I was later told was more than what a Vietnamese colonel earned in the same

Mama-San

month. For a warrior to earn less than a maid, what would that say to the soldier and the values he's fighting for? And if the mama-san is out-earning the colonel, how did that affect the millions of men below the rank of colonel? These men were ordered by their government into battle and expected by their officers to give their lives in mortal combat to protect not only a corrupt government in Saigon, but through extension, the earnings of a hooch maid. That seems a little skewed to me.

I'm sure the corruption was not planned, it just happened out of the situation as it developed. But we failed to recognize the problem, or just ignored it if, in fact, we did see. It was our ability to grossly overpay a hooch maid that was doing the Vietnamese society a moral injustice and lessening our ability to effect a positive outcome of that war effort. Our very mission to save the Vietnamese from the communists was doomed, in part, by our just "being there."

Wake Island

Shortly after New Year's of 1970 I was on my second flight across the Pacific Ocean. I had just completed a thirty-day leave provided to me courtesy of the U.S. Air Force as my reward for extending my tour of duty at Pleiku Air Base, Vietnam. I remember the flight starting at Tacoma, Washington, followed by landings at Hawaii, Wake Island and Guam before reaching our final destination at Cam Ranh Bay, Vietnam. I had left behind the snow and ice of an unusually cold Baltimore winter, and by the end of the island-hopping flight I was back in the blowtorch heat of Vietnam. I was not very happy to be back, but I had voluntarily extended my tour, and most likely, any disappointment I was feeling was related to the loss of dream-time at home and returning to my most recent reality, Vietnam. But, after that long-haul flight it was a relief to finally be at the end of the journey.

On the flight I had a window seat (bad choice when you have a small bladder) just behind the right wing. The landings in Hawaii and Guam were more or less conventional, but coming down to our second stop, Wake Island, was different enough to remember. To begin with, I had no idea we were landing at Wake Island after leaving Hawaii and I don't remember if we were informed. But I do remember there came a time in the flight after departing Honolulu where the conversation-killing noise of the jet engines suddenly decreased. The airplane then started endlessly circling and circling, but all the while descending. We were getting closer and closer with each revolution to all that blue water out there called the Pacific Ocean. As much as I craned my neck and flattened my face on the window, I could see nothing but water getting closer and closer as we were going down and down. Although nothing appeared seriously wrong, I continued to worry; but I did not ask "the question." No one asked "the question." As we got lower and lower the engines were almost silent. And below, the blue was becoming deeper and deeper. When we were within a few hundred feet of the ocean, the foreboding circling quickly became a straight path and out of nowhere a runway appeared under the wing. Seconds later the aircraft plopped down and came to a tire-smoking, reverse-thruster stop, thankfully before reaching the end of the runway. But talk about being in the middle of nowhere; well, we were now there. Welcome to Wake Island.

As we landed and taxied over to the parking area I could see little out the window; first, because my window wasn't all that big; and second, because there was little out there to see. After parking and securing the aircraft, we were allowed to get off and my first unrestricted view of the island verified my initial conclusion, "nothing here." It wasn't really empty or void of structures, as there were buildings and structures to support the aircraft operations and to sustain the workers, but little else. Among the structures was a small terminal/ snack-bar. After deplaning it was a pleasure to put my leg and back muscles into gear and to walk over to the building and get out of a suddenly damn hot sun. While the plane was being serviced and refueled, I had time to buy a soda and a couple post cards. I enjoyed the freedom to lull around in this beautiful nowhere, although I do remember that the mini R&R was becoming less attractive as the effects of the heat and humidity started to set in. Shortly it was downright uncomfortable. But soon enough, the airplane (I believe it to be a DC8) was ready to go again and we were all herded back to our seats. I was glad to see my camera on the seat cushion, especially since no one had thought to steal it to punish me for being stupid enough to leave it out as I had. I squeezed into a seat that was getting smaller and smaller all the time and buckled up for the next leg of the flight.

Leaving the little island was as scary as landing, maybe more. We rumbled out onto the runway and stopped. I imagine the pilots must have been standing on the brakes as they rammed the jet engines into full ear twisting thrust. We went nowhere for what seemed an eternity. "Are we overloaded?" I worried. The aircraft was shaking and making metal groaning noises from deep below my seat and then, with a slight lurch, we started rolling, but ever so slow. I was concerned because the roll was not in synch with the screaming engines, which told me we should be going faster. I swallowed hard. It seemed we rolled forever. Our acceleration increased to a really hot pace but the aircraft did not fly. We just kept rolling faster and faster but still on the ground. I swallowed again. The wings were bending ever higher when I felt a faint lift and we were flying. I'm sure the wheels were no higher than a few feet when the runway disappeared from under the wing. There was lift, a wisp of concrete, a blur of sand and then blue ocean. Fishing would have been easy at that point. I didn't like that take off, but then again, I guess any take off is good if you get airborne before running out of runway.

As we lifted off into that beautiful Pacific sky, it finally dawned on me where I'd been. I remembered that Wake Island was a place in the middle of nowhere where our fathers fought the Japanese early in the Pacific War. Men, brave men, had fought and lost their lives over this little crap island and all I could do while I was there was to think of my own butt, fretting about a landing and a takeoff that I'm sure the aircrews were expert at accomplishing. I remembered that the Japanese had executed close to 100 POWs rather than surrender them to the Americans when we came back to claim the island with a force of arms. It was a horrible fate for those people captured by the Japanese, and I knew that in this present war we had POWs suffering at the hands of the Viet Cong and other communist Vietnamese forces. It took a little while for the history of this fly-speck of an island to really settle in. And all I could do at first was to worry about the present safety my own rear-end. But when the enormity of it did sink in, I felt small.

Thing to Do Today

My ZIPPO lighter sits on my desk at home. Just recently I had the Zippo Company repair the lighter back to working condition. I had broken the flip top years ago when I was a two-pack-a-day smoker, but I never felt it was necessary to trash the lighter as it was a part of my personal history and I wanted to keep it.

I had purchased the lighter in the U.S. so I had it with me when I got to Nam in November 1968. During my time in Vietnam, I saw that many of the smokers had lighters that had been engraved with poems and artwork of one manner or the other. I decided that I liked what I saw and wanted to do the same with my lighter.

On one side of the flip top I had PLEIKU VIETNAM engraved and, in larger letters, MARYLAND was engraved on the body of the lighter. That work was done at a shop on the base run by a Korean concession. At a later date, when I was in Pleiku Town for a visit I thought it would be fun to have an X-rated engraving done on the back side of the lighter. There were several shops in town that offered that service and they used a mechanical pantograph machine to etch the lighter. All strokes were done by hand, no electricity was available to run the machine. The scene inscribed by the hand operated machine was of a couple having sex in the male superior Missionary position. Quite innocent by the standards of 2012, but very "racy" for 1969. Above the couple was engraved... THING TO DO TODAY... Man, did I ever own a cool lighter now!

I was apprehensive about sending my lighter to be repaired by Zippo. I put it off for years, but after watching a TV report on how Zippo reconditioned lighters, I decided it was time to send off this chip of personal history. I was well aware now about the excellent reputation the company had in standing by their lifetime guarantee... but I was attached to this lighter. I mailed it.

With great relief, my anxiety proved to be unfounded as the lighter was returned in a very reasonable time, well-repaired and ready to strike after adding some fluid. Of course I did not light it as by that time in my life I had quit

A photo for Lynn, 1969

smoking. It went on my desk. Just sitting there as my steel mechanical war buddy, a great piece of American engineering and manufacturing, just me and that ever-so-subtle reminder for me of my disappearing yesterdays, of the war and a wild life, and finally reminding me to count my blessing for the peaceful life I have had since coming home.

Cholon Dance

It was a night for drinking, dancing and whoring and I wasn't missing a minute of the action. I was in a crowd of rowdy dancers: GI's dancing with their girls; girls dancing with girls; or people too self-absorbed to even care if they were dancing with another person. The bargirls were numerous. Some enjoyed being unemployed for the moment and danced and laughed aloud. Several of the unescorted girls had already snagged a GI for the night, but by this late hour the troop was burned out or dead drunk and was most likely well-fucked and certainly passed out in his room, his money well spent. This, in turn, allowed the girl a chance to have a good time with her friends or even make some extra money by doing a "short time" with another GI just looking for a quickie. The music was as loud as the old jukebox could be cranked; it was all-American and non-stop. The beat was incessant, the dancing frenzied and the scene was a needed relief for guys burdened by the boredom or frightful horrors of war. This was a night for forgetting the war and the music; booze and sex helped – it helped a lot.

I was in Saigon and it was the summer of 1969. Through open windows and balcony French doors there came a slight breeze. But the thick, warm tropical air was not sufficient to stem the ever-rising tide of sweat, especially among the uniformed and booted GI's. The windows and balconies were covered with chain link fencing, giving an even odder appearance to the old French-style hotel, but it did add a bit of protection from any hand grenade tossing friend of Uncle Ho. But that night was neither the time nor the place to be concerned with something so trifle as a hand-grenade. Unrestricted drinking and unfettered women have that war-numbing effect on testosterone-bathed brains: fuck the grenades and the Charlie who chucks them! As all the others on the floor that evening I was sweating, drinking and doing a dance of abandoned forgetfulness.

I had arrived at the Hanh Long Hotel in Cholon two days ago, and this was my last night before returning to the air base in the Central Highlands at Pleiku. Cholon was the ethnic Chinese quarter of Saigon and the hotel had

Jerry and Maria, 1969

been recommended to me by a buddy back at the air base. I had earned a three day pass and I decided to take his recommendation and fly down to Saigon for some in-country R&R. Of course, R&R was supposed to be military shorthand for Rest And Recuperation but in the "GI speak" of the day it translated to Rape And Rampage. Now this is not to say that there would be literal rape, but just lots and lots of non-stop sex with as many women as you could afford. And the Rampage was more of a controlled rage brought on by excessive booze, plenty of women and a "not give a shit" attitude fostered by just being in Vietnam. Of course, some guys would "go off," but for the most part these young men were restrained or protected by their buddies or quickly subdued by the military police before any non-repairable situations could arise.

My 170-pound frame and the gyrations of dancing kept the booze from taking full effect, maybe the Jack Daniels was watered down, or a combination of it all; but I was still vertical and having one helluva time on the dance floor. The psychedelic music coming out of America in 1969 complemented all the craziness of the surreal war in Vietnam that held me that night, and I just couldn't get enough of the insanity to please me. Even the openly-smoked joints couldn't numb the growing panic of being. The smoke from the Marlboros, Salems and the joints was heavy in the air, suspended by the thick dampness from the just-passed thunderstorm. During the rainy season monster thunderstorms cooked up rapidly, day and night, leaving the air wet, very wet, even after the torrential downpours stopped. It was the summer of 1969 in the Hanh Long Hotel bar, on the seventh story top-of-the-mark, in the war-bloated city of Saigon. It was the perfect time and the perfect place for the perfect insanity of my 22nd year.

But it was about to get crazier, crazier beyond my ability to rationalize or begin to understand. I had just slugged down another Jack Daniels, in a line of uncountable drinks, and was going like hell on the dance floor. I have no memory of who I was dancing with if, in fact, I was dancing with anyone in particular since the insane shaking frenzies required no touching, no looking, nothing but your own self-absorbed wildness. Deep into the night, before we were all completely dead to reality, somehow through my glazed eyes I saw huge flashes of light outside, beyond the city limits, miles away from my debauchery. Although I was almost completely brain-fried, I could see the flashes plainly and at first I wanted to dismiss the display as lightning, nothing more, and nothing else. But in a sobering thought I realized that God-made lightning did not come with tracer rounds. Hell red tracer rounds by the thousands arched skyward, burning their way through the fetid night and up into the heavens above, accentuating the explosive flashes of what I now understood to be the curse of man-made lightning, war.

I stopped existing on the dance floor, transfixed by what I was watching. All was quiet. There was nothing going on in that room that I was aware of any longer. There was no music, I couldn't hear. There was no whiskey, I was sober. There was no smell of sex, I was celibate. Only the far off show of destruction and terror held my attention and my focus was on the sudden realization of overwhelming death that was also busy enjoying that gruesome evening. I didn't understand why, but I felt the need to pray, something wasn't right.

I Hate the War

The rain announced itself with little splatters and bouncing raindrops, here, there….one and ….another and another endlessly piling on

The popping on the rusted tin roof becomes a staccato, hammering song as the first waves of serious rain follows the first calling wetness.

Steady…steady, soft rain, laughing happy rain clearing the air, washing the streets and I wonder where you are

My heart is beating and I can hear it here in my dark room

I am naked except loose shorts and my cigarette smoke rises and then falls with the heaviness of the air….I drag deep on the cigarette and flip it into the night to be washed away in the street, 3 walk up floors below

I have been waiting for you, waiting to be once again inside the woman you are…..deep inside of you taking you hard as a man who is about to explode

But it rains and rains on……steady and I see the drops suspended in the air as they fall past the only street lamp near to me

The curtain flies back into the room and my bed netting waves with the first big slap of thunderstorm air shooting into the room

A distant rumble comes and rolls along and I am misted with Sai Gon rain, night air and a desire that makes me hard…

The shop windows below are 9 pm bright, hawking their wares in garish neon displays but no one walks past…life is resting, but not me

My nipples are erect from the slight thunderstorm chill in the air and I hurt for you, I am addicted to your body and a prisoner of your soul……….and you know it, you know it all too well

And so I wait…the thunder rolls ever closer and still no sign of you….no smell of you, no smell of us……

Lightening streaks the air in a beauty second to none, but all I want is to taste your passion sweat……I light another cigarette and I wait: I hate the war

Part II: Some Tales from R&R

Kilroy Wasn't Here

Graffiti, scratched or penned here, there and everywhere, is a continuing time-honored practice of soldiers who feel it necessary to impart the wisdom of their time. Some of the writings are designed for stunning effect, but by reading the message it is clear most were not, probably most are not. My first introduction to GI graffiti as a kid was the ubiquitous "KILROY WAS HERE." It was particular to my parent's World War II generation, and at a time when the "Big One" was a fresh memory in everyone's mind and the vets were still vibrant men, our dads. I knew KILROY so well that I even scribbled my own rough interpretation of the well-known drawing with lumps of coal on my canvas, the nearest sidewalk or wall.

In Vietnam, my war, there was always a place for a GI to sign in. The most common sights included the whorehouses and latrines, where hundreds of jokes were secretly scratched out each day. There were sick jokes and VD jokes. There were tasteless two-line poems and ditties. There were jokes I didn't understand, always the overwritten dick and pussy jokes and, of course, the even more tasteless racial jokes written by everyone about everyone else. Occasionally, there was a really funny gut-buster that would cause an outburst of laughter even during the most trying bowel movement. But the point is, little did it matter where the GI was or what he was doing, he always found a bulletin board of his own making to scrawl his neo-Shakespearean prose or dirty little joke.

I had departed Pleiku Air Base in the Central Highlands to partake of the liquid and flesh available in Saigon. On the advice of my buddies who proceeded me, I was staying at an old French period hotel, the Han-Long. The building was a nondescript, plaster-coated something or the other that had suffered from the lack of paint for many years. All the windows were protected with iron grates and chain link fencing. Inside, slow moving fans wobbled overhead from all the ceilings and in the small rooms that surrounded a central open stairway. The somewhat stale rooms came standard with western toilets, which were a welcomed relief from the gross French-style hole-in-the-floor shitters that were so common in Vietnam. And, of course, there were the ever-present tropical roaches. Big ones. Apart from all this luxury, the real attraction of the Han Long was the bar on the top floor. Perched atop the seventh floor it was a bar as tacky as they came, centered on a large jukebox. The U-shaped bar was well stocked with all the top brands. The black market booze looked and tasted good because it was the best stuff that could be stolen off of the American bases. The old mama-san who tended to the cleanliness of the floor sold marijuana in Salem cigarette packages. Tasty buffalo burgers were cooked to order with lettuce, tomato, mayonnaise, and onions. But, best of all, the booths along the walls were well stocked with attractive women who were more than eager to meet the foreigners in their country. Of course, the proper introduction needed to be made, and that usually involved a quick haggle for services and prices, followed by a quick exchange of GI money called Military Payment Certificates (MPC) or, occasionally, the illegal U.S. greenbacks. The greenback always got a better exchange rate as it could get back to the Viet Cong faster. But for now, who cared, twenty dollars was twenty dollars and it purchased twenty-four hours of companionship and R&R bliss.

It was, in one of those booths, in the bar, in the Han Long that I selected my personal escort for the next forty-eight hours. She was a fairly young girl with the ubiquitous name of "KIM" who was equipped with standard black hair, black eyes, fair skin and English language command sufficient for conversation.

On the second afternoon of our companionship, Kim took me home with her. It seemed she needed something or the other, and I was up for the adventure. The short horn-blowing, near bicycle-crushing cab ride led to an alleyway (and, I suspect, the continued employment of the taxi driver with a kick-back) where we stopped, I paid. We continued by foot down the wet dirt alley and, within a short distance, we stepped into the ground floor entrance of a three story walk-up. The coolness of the concrete caught me by surprise, but it was a welcomed relief from the heat of the early afternoon. We ascended the dark narrow steps with Kim in the lead, me following with my hands on her pleasant bottom. On the second floor a raucous card game was under way in the hall. There were numerous old ladies slapping small cards hither and heather upon the floor. Kim spoke to one of the old ladies (my introduction? Most likely not). She stopped her game for a second and glanced up. She caught my eye, flashed a quick beetle nut smile and returned to her game, slapping another card on the floor. Nice to meet you, too.

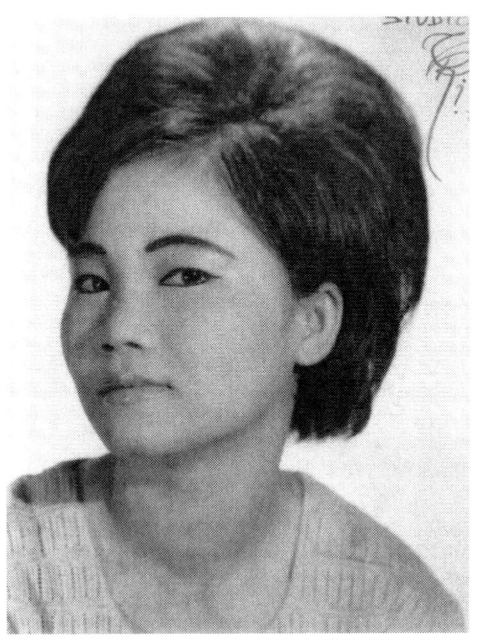

Kim of the Han Long

Kim and I stepped beyond the card game and down the hall towards the front of the building. Within a few feet we were in Kim's room. It was small but airy and punctuated by a balcony and another large window where Kim's sister sat in front of a mirror, fussing with make-up. Kim spoke to her sibling and ignored me. She held a racing conversation, sharing secrets of whatever, and I'm sure they were coaching one another in the art of applying too much make-up, especially that really gooey stuff on and about the eyes. The corner room was flooded with light, the sheer curtains were fraying in the steady cross breeze. All in all it was not uncomfortable, I actually found it… pleasing. After soaking in the room's ambiance, I noticed a GI quietly sleeping on the sister's bed, mouth open, breathing easy. The mosquito netting down so he breathed no flies. His green tee-shirt was somewhat sweaty and his feet were shorn of the green socks and jungle boots that we all wore. His uniform was dirty, field dirty, and I assumed Kim's sister was his "steady" by the casualness of the situation. There was only one conversation in the room and I was not part of it, so I flopped on the empty bed to relax and maybe lull-a-bye the afternoon myself.

What I assumed to be Kim's bed was hard against the wall, and at first I thought the wall was just typically Vietnam dirty and paid little attention to it. However, being in the horizontal position, up close and personal with the wall, I realized the dirt to be hundreds of signatures, poems and whatever, in the hand of the English language as I knew it. Doubtless, it was the most fascinating and amusing thing I ever saw. Here I was perhaps GI number one thousand, two hundred and sixty-nine whose ass had flopped on this bed, that had run the gauntlet of card players, who had climbed the concrete stairs, who had employed a taxi driver for nothing better than a short walk, and who had employed Kim. Damn! This was really neat in my way of thinking. I read with unbridled joy the wisdom of the wall, the nave novel of GI authors, calling card beyond prior experience or belief.

Kim finished her conversation and beauty treatments. Her sister looked at me for the first time through her heavily coated eyelids, licked her lips and laughed a laugh of carnal knowledge I thought I understood. With a "Hi Ho Silver, Away," Kim and I were off for, this time, a short walk to the Han Long, the bar, and some more GI R&R.

The Hong Kong Smile

The small girl directly approached me, curious for a child of her years, maybe five at most, and highly unusual for a Chinese child. Her straight black hair was cut in what I observed and assumed to be the usual fashion. She was a real cutie, dressed in street clothing without need for mending or washing, and she appeared well nourished. She was clean, she smiled, and she actually spoke English to me.

I was a GI on a quick Rest and Recreation (R'&'R) up to Hong Kong from my air base in the Central Highlands of Vietnam. Every week the base bird, an ancient World War II C-47 cargo plane, flew from Pleiku Air Base to Hong Kong to pick up various critical electronic parts. For the most part there was enough space on the aircraft to take along five or six additional men on each flight and the trip was one that was prized among the airmen on the base. I don't remember how it came to be that I got on the flight, but I did. We were scheduled to stay only three nights and return on the morning of the fourth day with the necessary supplies safely stowed in the aircraft.

Following our early morning take off into the rising sun, the flight to Hong Kong was uncomfortable and cold, and it seemed as though the flight took forever. The lumbering C-47 operated near 10,000 feet and that, for us, was bad news – the airplane was not insulated or heated. To men still in tropical fatigues, the trip became quite chilly early on. There were several hard metal chairs along the fuselage. But they were too hard and it was too damn cold to sit there for any length of time. The only comfortable accommodations were the two medical stretchers attached to the ribs of the airplane which were quickly claimed by the senior sergeants. My final choice for the flight was a bed of parachute packs that were, surprisingly, hard as rocks.

After five hours of flying we finally thumped to a landing at Hong Kong's Kai Tek airport. The aircraft was parked in a remote ramp area, as required for its protection. Once the props stopped turning, we were all unloaded, and the enlisted men were segregated out and piled into a pre-arranged minibus. Our destination was the President Hotel in Kowloon, along the shopping mecca of

Nathan Road. Our Chinese host settled us into the hotel but soon reassembled the group to speed us off to do some shopping, followed by the last stop at the California Bar.

Without details of the story, the California Bar is where I met Jenny. My story with Jenny is the same story as most GI's so far away from home. For a modest price she became my tour guide, shopping director, and bunk-buddy. This was a most convenient "all inclusive" package. What more could a twenty-two-year-old GI ask for? As part of the package deal, Jenny took me up to the Tiger Balm Gardens. It was not a hot spot for tourists, but it was clean and relaxing and it offered scenes employing some of the most grotesque sculptures and painting that I had ever seen. It made for great pictures to send back to the folks at home!

Upon taking a break from climbing up stairs and rocks, I was high enough to look across the eight acre gardens and further on to the next hillside where untold thousands of scrap wood, cardboard and tin shacks where innumerable people lived. I learned from Jenny that these people were illegal immigrants

My Jenny

Hong Kong hillside slums, 1969

from the People's Republic of China who had unlawfully entered Hong Kong where they wanted to stake out a piece of the British Colony's prosperity. Although these people were literally on top of one another, they were free. As Jenny continued to tell me the story of these hillside people, I vividly remembered watching houses such as these wash down the hillside as recreated by Hollywood in the movie "The World of Suzie Wong." I could see the dashing actor Bill Holden, as Robert, fighting the torrents of rain as he and Suzie (Nancy Kwan) desperately sought out Suzie's baby, Winston, who was in care of the Ah-Mah, his baby-sitter. Jenny's touch ended the scene and I returned to the reality of the moment.

As we sat there, the child boldly strode up as though out of nowhere. Looking directly at me in silence, which was strangely unsettling, she captured a fresh breath and spoke, "You want to take my picture?" How could I refuse? I was flattered, pleased and surprised. She was carrying two twigs in her hands and within an instant she struck a pose and directed her gaze at my lens. I focused, matched the needles of the light meter and eased down on the Nik-

The Tiger Balm model

kormat's shutter release. The camera's mirror flopped, the shutter snapped and the scene captured. I was happy. As I brought the camera down from my eye her little hand shot out, palm up, and she announced, "Fifty cents."

I suspect this child of the Tiger Balm Gardens is now a grown woman of forty-five years or a little more today. She may very well be a mother herself with a bold child of her own. It was just for a minute that this child was in my life, but her face will evermore be in my mind, as long as my memory survives.

Chungking Mansions

Guests at the Chungking Mansions hotel hostel complained about a bad smell coming from a room. Staff called the cops, who in turn called the fire department. When they broke in they found two bodies that had been dead for two days. Swell. And yes, with the heat we have been having lately, the pun is definitely intended. Chungking Mansions has a history of violence and is regarded as a firetrap.

WOW! When I was reading this post on the internet back in April of 2002, I was somewhat stunned, for I had stayed there before.

The posting continued and gave a description of the place:

"The Mansions" is a shuddering collection of some 920 guest-houses, hostels, shops, restaurants, flats and factories crammed into an aging 17-storey tenement. By reputation it's a place where you can satisfy almost any vice. Residents say drugs and prostitution are today's problems, but the Mansions' real notoriety stems from its days as the center of a gold-smuggling trade to Nepal.

I read the entire web posting as the article provided a sordid fascination, and I knew the place. So, to the web posting I added a memory or two to include my "adventures" in the Mansions when I was in Hong Kong in May 1969 on a three day pass from Pleiku. However, I just found it so amazing that I could trip over a link to my past in this strange place of the internet.

I added my story about my overnight stay at the Chungking Mansions. It lacked real detail since it was not meant to be a sex story of a Vietnam GI, but just a passing thought or two about my association with the subject of the posting. My posting was:

Boy, sweet memories! Been there, done that back in 1969 as a GI on R&R out of Nam. Stayed there with a pretty girl named Jenny.

I remember looking out onto Nathan Road and loving every gaudy neon minute of the scene. Crappy little room, with guy stationed at a desk out in the hall, near the elevator. Had some tea and didn't mind the smelly little bed with semi-clean sheets, as there was something else on my mind.

In my memorabilia I found a business card I had retained from my visit. I still have it and it reads:

ChungKing House De Luxe Hotel, ChungKing Arcade Block, 4th & 5th Floors, 40 Nathan Road Kowloon, Hong Kong. WHERE TO STAY WHILE IN HONG KONG. Tourist Shopping Centre Most Reasonable Rates. Fully Air Conditioned, Private Bath, Telephone, Radio and TV on request. Cable Address "CHUNGHOUSE"

Oh, I forgot to add that it was a large business card. And all of this was also on the other side of the card, in Chinese.

Now, I don't remember what the reasonable rates were, but I do know I had to stay there as the hotel I was booked in, The President (owned by Hyatt) would not allow "working ladies" onto the premises. So, the ChungKing was the closest stop with a bed for rent for those, like me, not interested in sleeping. Worked for me!

Part III: Coming Home

My First Kiss I Remember

I don't remember my last day at Pleiku Air Base. I know there was a last day but I don't remember the specific day or date nor anything associated with it in particular. I do know it was July 4 when I arrived back in Baltimore, so most likely I left Pleiku on the 29th or 30th of June 1970.

My first full day at Pleiku was Thanksgiving Day of 1968. It was after arriving in Vietnam, sleeping a night on the concrete floor at Cam Rahn Bay passenger facility, flying up to Pleiku in a God-awful, ear-bursting C-123, and then spending another night in the transit barracks at Pleiku Air Base. The night of my arrival there was an attack on a nearby army camp off to the west. The mortar and rocket attack was very visible from high atop the airbase cantonment area where the barracks was located. I didn't know the camp's name then, and I don't remember it now, but I do remember watching that attack in morbid fascination. And I do remember very well the sight of the fire from the orbiting C-47 "Spooky" working over an area near the army camp that was still getting plastered. Those memories are as though it all happened yesterday. But I don't remember the day I left Pleiku nineteen months later. I wonder why.

I don't remember if it was a C-130 or a C-7 that flew my butt down to Cam Rahn Bay the day I departed Pleiku. I don't remember anyone giving me a ride down to the passenger facility or even saying good-bye. It's just not there now. I don't remember arriving at Cam Ranh Bay, but I remember bunking at the transient barracks, the Cam Ranh Hilton. I lingered around the sprawling airbase for a couple of nights, just wasting time. The little buses operated by the Air Force were all over the place and running a regular schedule, and I thought it great fun to hop on and ride, wasting time. But all the while as I rode, I struggled with a strange emptiness inside me about going home. Or was the emptiness there because I was leaving? I was confused.

I should have been overjoyed about going home, but at age 23, with over a year and a half across the big pond, I wasn't really sure where home was anymore. The intensity of Vietnam, the friendships, morbid adventure, the travel, sex and drugs, the "otherworld" magic and corruption had confused me. I knew very well where I was born and raised, but I wasn't sure that that physical place was home now. But if it wasn't, where was home? In my confusion, I still felt I had a duty to report back to my parents, to the people who loved me and worried about my individual health and welfare. At that time in my life no one else cared a hoot for me save the people who brought me forth on this earth. I figured I owed it to them to show up. If anything was going to make me go back to Baltimore, it was love, but I was still confused.

As I wandered around that huge base I remember buying several souvenir items at the big BX. The little items I don't remember, but my purchases included a book, *Great Victory, Great Task* by the North Vietnamese Commanding General Vo Nguyen Giap. I thought it would be interesting to read a book by the guy who screwed-the-pooch of the French colonialists in an earlier war and was beating the hell out of America's politicians by losing and continuing to lose on the battlefield. I still have the book. I read it once and put it away.

I also found the barbershop and got all my hair cut off. I figured it was an insurance policy against any chicken-shit lifer who wanted to add a measure of salt to my last days in the service. I just didn't want Sergeant Dickhead or maybe Lieutenant Asshole stopping me to give me a raft of shit about the length of my hair. Damn, I knew my hair would grow back no matter how short I got it buzzed.

When the day came to actually board the "Freedom Bird" home, I guess the departure processing, boarding, and seating just didn't measure up to a moment to be remembered. I don't remember the flight from Vietnam back to McChord Air Force Base. I do remember a lot of whooping and applause when the aircraft rotated off the runway, but the rest of that long, cold, boring flight was also nothing to remember, and I guess I haven't.

Leaving on a jet plane

I do remember very well our deplaning at McChord in the very early morning. There were reporters there with cameras and very bright lights. I didn't know who they were then and I don't remember now if I was ever told. It was predawn dark and the movie lights were dazzling, way too damn bright and too close for comfort. I responded to the stimulus with a very stiff-armed, middle finger salute. For some innate reason that salute became the appropriate thing to do at the moment and today brings a smile to my face as a fond memory.

Inside the reception area I was briefly questioned as to the contents of my luggage. I believe the Air Police fellow, or whoever he was, asked me about having possession of any illegal drugs and pornography (all of which I left at Pleiku) and my reply was a truthful negative. He then passed me through without so much as a wasted peek in my bags. I don't remember processing out and receiving my separation papers from the U.S. Air Force except for the several $100 bills I got from the paymaster people. Those bills with Mr. Franklin's face on them always made an impression on me. They were nice and crisp and something to remember, just like my first kiss: her name was Vicki.

... And Then I Was Home

July 4, 1970, was hot and sticky in Baltimore. It was the kind of weather that made the local people thank God for the Orioles, blue crabs and cold Baltimore-brewed beer. That was my town and a simple way of enjoying any sultry July. And it was the day I got home from Vietnam.

My flight home had departed Seattle-Tacoma Airport early in the morning. I was the last person to board the plane with my standby ticket, and, as surprises tend to be sweet, I found myself seated in the last available seat on the plane – in first class. Now that was sweet. But I was lost in first class. Everyone was dressed so nice, unlike today when being a slob is the rule, and I was the only one in uniform. Although I stood out, these first class people did not regard me with aloofness or disdain. But no matter how nice they were, I felt uncomfortable. These were people with money and I was but an unemployed GI on his way home from Nam. I was envious of their positions and their abilities to pay first class and I didn't fully take advantage of all the perks because I did not know what the perks were in that part of the cabin. I just sat back and minded my business and was polite enough to say, "Yes, thank you," to the very pretty stewardess every time she asked me if I wanted this or that... and I did, no matter what the offer was.

Although the flight from the Seattle-Tacoma airport was almost six hours long, it appeared to be in no time we were landing at Baltimore's Friendship Airport. The airport's pleasant name went along with its small city location. Soon enough, I was departing the comfortable seat and seeing for the last time in my life the so very pleasant smile of the beautiful, strawberry-blonde stewardess. She was the first class angel who offered me blankets, pillows, drinks, food, more drinks, matches (back in the prehistoric days when I still smoked and did so without thinking) and the sweet softness of her voice. Although I knew she had been well trained to do all this for the customers, she was so damn good at it I could have married her right then and there. But I filed out of the plane like everyone else.

There was a rather large crowd of people waiting for the deplaning passengers as it was July 4, but there was no one waiting for me. My fiancé had

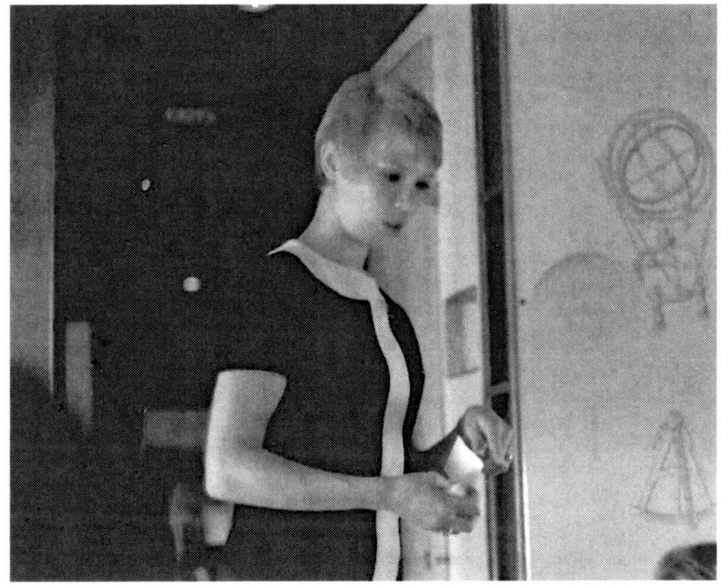

Homeward Bound, 1970

dumped me a la "Dear John" and neither of my parents drove, never did. On occasions in the past, a neighbor would give my parents a ride to the airport to meet me when I came home and save me the taxi fare or the ordeal of hitching over to the east side of Baltimore where I lived at the time.

I made my way through the crowd of strangers at the gate and down to the baggage carousel where I waited for my AWOL and duffel bags to be dumped out. As I stood there I lit another Salem and smoked thoughtlessly, just waiting... puffing. Two days ago I was in Vietnam. The bags started spilling out, and it wasn't long before my bags arrived and I collected them with but a small fight from a little old lady elbowing me for a place at the baggage claim.

The cab ride home was made longer by the heat and humidity of the day. The cabbie stopped right in front of my parents' row house, and before I got to the door, it was opened and there they stood, waiting, strangers. After all the hugging and kissing stuff that made me uncomfortable at the time, I climbed the bare steps to my room. At least it was the room that I used to call mine, and I felt uneasy. It was as though something should happen, but it didn't. With a cautious unease I changed out of my sweaty uniform, showered, and put

1967 Chevy Camaro

on something a little more civilian, which included my old high school Jesus sandals, strange.

My Chevy was out back, freshly washed by my dad as his way of saying welcome home. The car started with no problem and I drove away; for, in truth, I had nothing more to say to my parents, not now. It was as though I wasn't really home. I didn't know what home was and I just felt utterly lost not being in Pleiku.

CHAPTER TWO
After the Storm

The first story in this chapter, though very short, is also very significant to me as it was an event that illustrated how some things never go away, no matter how deep the memory is buried in your mind and soul. And it is also a praising of how one can find peace and quiet... after the storm.

The other tales are thoughts on what happens day-to-day and how small things can turn us around, if only for a minute; and how a movie personality can earn the eternal loathing and damnation of a generation of men who served in Vietnam... from my perspective.

Of course there are many things to write about, but these are highlights, flashes of time now gone, and these memories are fading with age. So, I write.

A Thunderstorm in July

Nothing had ever happened to me to give me the really intense "flash-backs" that I hear being described by Vietnam vets. Maybe once in a blue moon I'd smell something really rotten and remember fondly when I fell into the sewage ditch at the bottom of the air base cantonment area, but that's another story.

Earlier this month we had a thunderstorm in the middle of the night... short, but violent. Normally, I love thunderstorms – the sound and fury, the pure, raw power of nature. It was the dead of the night, the point where one is in that deepest of deep sleep, when the first round came in.

The burst of light and the slapping concussion of the near miss set off a car alarm. I was shocked awake and quickly heading for the bedroom door in a crouch. The rain was pounding. My wife asked me, "Jerry, where are you going?" "The bunker," I answered.

I was not afraid. I was not shaking but I was there once again and the rain of rockets made me wish for something... something I couldn't figure out.

I straightened up, walked over to the window to look out into the blackness of the night and to check for blowing rain. No rain was coming in, good. There was nothing I could do about the car alarm. I flopped back in the bed and snuggled with my wife of twenty seven years; I thanked God quickly and said a prayer for those who did not return.

Summer Class, '99

In that summer I was attending a class at a local community college in an effort to keep the old gray matter working. Before class one evening I was discussing the bombing of the Chinese Embassy by the U.S. Air Force. In response to the bombing, Chinese students had arisen in Beijing and were gleefully stoning the hell out of the U.S. Embassy.

It was my contention that the leaders of the People's Republic of China were manipulating the whole affair. They had closed the schools, provided free transportation and box lunches for the students, and supplied all the stones they could chuck. To me this seemed as though the entire ongoing "demonstration" was great sport and a real lark for the students. What undergraduate wouldn't take his place in the stone-throwing crowd? No class or lectures; lunch on Mao's tab; and best of all, fuckin' up the American Embassy. Fun!

The lady I was speaking with was about my age and she thought briefly about what I had just suggested. "Yes," she said, " It's just like when we used to throw stones at the soldiers on campus during the Vietnam War!"

I turned in my seat and went back to minding my own business and never spoke to her again. Maybe that was little of me, but it was my reaction. It's funny how this shit can come out of the woodwork when you least expect it. Was the stone throwing a lark for her, too?

Father's Day and Jane Fonda

Father's Day was a strange day for me to be thinking of Jane Fonda, but I did. I cannot decide, and I'm not really qualified to decide, if Hanoi Jane is a "new woman," a better person, or a saved soul now that she has found religion. However, I do know that she was very visible and very active during the war and that her anti-war, pro-Ho antics cost the lives of American men in Vietnam. She is directly responsible for making widows and orphans and cheating those men of Father's Day.

Perhaps after all these years that's a lot of baggage for Jane to be carrying, and maybe that's why she asked for help and turned to the strength of God for that assistance. I know that for me, at age 54, I'm fairly set in my mind and that I'll never forgive the bitch. In her search for personal redemption, I'd like to see some very public good works done by her. Perhaps that would help with my negative attitude. I'd like to see her turn some of her fortune into charities to help widows and orphans of the war in Vietnam; I'd like to see her doing good works in many areas but, I will not hold my breath. Maybe I'm a little too cynical, but I think the current story of Hanoi Jane right now is the same story of her last five decades... me, me, me. It's all about her and what she wants, not how she can help.

For the most part, I won't really give a hoot about her conversion. I'll just enjoy my children and my grandchildren and, when I'm with them, I'll most likely not give Jane a gnat's fart of a thought. However, as I listen to my grandchildren giggle and laugh, as I make a clown out of myself for their amusement, I will remember to thank God for that pleasure.

When my house grows silent and my children and grandchildren have gone home, maybe I'll hate her just a little more knowing that it was because of Hanoi Jane that this singular pleasure of a child's laughter was denied to so many Americans. Jane, you should be ashamed.

Aw Boon Haw Gardens

I read some time ago about the anticipated demise of the Aw Boon Haw Gardens in Hong Kong. It is a place I knew as The Tiger Balm Gardens. The plan, as I understand it, is to use the hillside site to provide the vertical space needed for additional high-rise apartment buildings. I'm sure that the housing is needed and that the land is valuable, much too valuable for maintaining a relic of the past. I realize, too, that there is little in Hong Kong that any historian would call an historic relic. The government tears down old stuff with few exceptions. There is a desperate need for any horizontal space to construct the apartment towers. However, this is personal for me as those gardens and the Tiger Pagoda are a part of my life and my personal history. It makes me sad to know that this amazing place will be dozed into memory.

I first came to Hong Kong in November 1969 as an American GI on leave from Vietnam. I didn't stay long and my behavior, I regret to say, was not the most appropriate for a visitor, but then again, who expected "proper behavior" from a GI on leave? But I fell in love with the city and the experience and by the time I departed Hong Kong would forever be in my mind and my heart. It's been well over forty years since that visit and I'm pleased that I still have vivid memories of what I remember as the Tiger Balm Gardens. Visiting the gardens brought a strange and unexpected peace to my soul amid the vividly painted grottos and pavilions that display concrete visions from Chinese mythology and a very garish area depicting the horrors of hell awaiting evil men. It was in these strange gardens, under the blue sky and warm sun, that the lovely Jenny, my Chinese bar girl companion, just sat and talked to me of nothing special. She made me laugh and then kissed me. There was tenderness to her touch and her kiss that made me feel human. Somehow, she made me feel saved.

I have memories of Hong Kong, and to this day I continue to visit the Island. I am sad that the Tiger Balm gardens are to disappear. I confess fully that I'm in love with Hong Kong for the peace and humanity I found in the Tiger Balm Gardens and the memory of Jenny's kiss.

Your Name I Cannot Remember

There are connections in this life that cannot be known as they are happening and are hard to understand once they have happened, and this story starts with a very rainy day in June 2005. It was the first week of June and I made it a point to drive across Baltimore to visit a person I once thought of as a friend from my early teen years.

It was raining like hell that day, but the thought of meeting Dennis again was worth the ugly drive. I had last seen Dennis in 1969 at, of all places on this earth, Pleiku Air Base in Vietnam. At that time Dennis had completed his enlistment in the Army and was once again a civilian. He had served with the 4th Infantry in the Pleiku area as an interpreter with the intelligence folks. He was a smart guy and learning the language of Vietnam had been easy for him.

Dennis had returned to Vietnam to be an entertainer, doing the magic act he had been working on for years and years, and to continue his courtship of a young Vietnamese girl. Dennis later married Lee, the only love of his life, and they would travel life with their own levels of success and failure.

That day when I arrived at Denny's Magic Shop the hard rain had become as hard as the monsoon sheets that I remembered from Vietnam. The steady wind pushed the rain into wet stinging beads as I ran across the crumbling parking lot to the magic shop. Shaking the wet from my hat and brushing at my shoulders I made my way into the musty shop to be greeted by his very wide entertainer's grin and the irritated grunts of a huge Vietnamese pot-belly pig.

Dennis was dressed in the disheveled style of an obvious sixty-year-old bachelor. His shocking red hair had morphed to mud and gray and his head seemed to be smoldering, which I figured was from his chain-smoking Pall Malls. We shook hands and, again, he popped a cigarette-yellowed showman's grin.

During my visit with Dennis, we talked and laughed about the past. I even endured his stories of heartbreak and divorce and continuing love for Lee. Then he turned the page in his head and brought me up-to-date with some of the old gang. Dennis had kept in contact with most of the guys, fellows I left behind even before going to Vietnam in 1968. Dennis approached the upcoming history lesson with a fuss of finding his lighter and then laid out the story as he fired up another Pall Mall. Several of the old gang were dead; drug abuse and AIDS had claimed one, a nasty cancer another, alcoholism had ruined yet another life, and sadly, even some of the children of my old friends had died. These were truly sad stories and they forced me to remember the laughing faces of teenage boys who were now either dead or human wrecks. I hated the pictures I was creating.

The group of guys I hung out with at age 14-15 were all older than me, except one. This caused me some pecking order problems, but it also taught me lessons before my time such as drinking, smoking and even the loss of my virginity thirteen days after my fifteenth birthday. As losing my virginity to Doris in the Buick was being re-filed in my head, the conversation with Dennis continued as he asked me if I remembered Tommy Taylor. With a slow search my brain turned up a faded face to go with the name. Yes, I remembered him. In fact, as the mental file continued to open, I specifically recalled that we use to call him Tommy Tucker the Chicken Fucker. Now, as to why we hung that name on him I did not remember, nor was there an asterisk in my brain next to his nickname. Who knows? We were smart-ass street kids and I guess that was just our basic callousness and crudeness. Tommy was younger than me, but he was tolerated by all because of his older sister. She was a gal we all lusted for, and just the sight of her had us all stumbling around stupid.

As this conversation developed with Dennis, try as I might, Tommy's sister's name I could not recall. I then relied on Dennis' memory but he too was shooting mental blanks. So, for the remainder of this tale I will call her "Darla."

As the magician continued with the story of Tommy he came to the announcement that The Chicken Fucker (Tommy) had been a hero of Hamburger Hill and had, in fact, received a Silver Star for his combat actions. I was a little disconnected by the thought of Tommy Tucker being a hero, but I also felt a certain pride in knowing someone from the old neighborhood had given better than he got in The Nam.

We talked a little more about Tommy and the war, but in just over an hour our conversation grew as stale as the cigarette smoke and I ran out of the mental gas necessary for the continued road trip along a muddy memory lane. I mean, I liked Dennis, but if the magician had made himself disappear at that moment, it would have been no loss. I hung out a little longer, fussing with the books and everything else there about MAGIC and then realized that it was past time to shake his hand and wish him farewell.

Again I dashed through the rain to get back to my car, escaping the moldering past and the remains of a man and a time I once shared with him and others. Those memories were a compost pile now, and I was also glad to be rid of his pet pig – disgusting.

That evening, with my brain back in normal mode, I got to thinking about Tommy and, as part of that process, I thought about his sister as well. In fact, I shamelessly considered his sister more than I thought about his bravery and heroism. I was proud of him, but his sister… well, she was simply more enjoyable to think about.

My memory of Darla was groovy (1968 speak). She was a goddess of heroic proportions, thin but meaty solid, looking so sexy in tight shorts that I could truly picture her bare ass in my best Technicolor imagination. And above the waist were those perky, medium, untouchable breasts. She was fair of face with dirty blonde hair, but what made her so special was that "look of sex" about her. It is that look that men recognize and love and women recognize and hate. Darla was not a stunner, more a plain Jane in truth, but a woman who had, again, that special unknowable that any man would sacrifice a testicle for just to kiss her. And Darla's obvious self confidence came easily as she fully realized the power that she had over men… especially us stumbling boys.

By the time I was in the Air Force a few years later Darla was working at a bar and dance club called The Hollywood Palace. It was a night club located at the Back River Bridge and, at one time, had been a rather fancy night club for the locals of twenty to thirty years earlier. There had been Big Bands to swing to, but that was another time, another war. In 1968, the Hollywood Palace was seedy enough to be cool for this new generation and sufficiently large enough to allow the bands to whip up several hundred frenzied psychedelic revelers. Marijuana smoke saturated the atmosphere as well as the earthy smell of beer, hard liquor, and the occasional whiff of urine from overused and under-cleaned toilets. As the bands hammered out such tunes as WAR and In-A-Gadda-Da-Vida, in the smoky club darkness, there was "near sex," but actual "balling" was reserved for the car in the parking lot.

Although Darla had not seen me for years, she remembered me as one of the boys from the old neighborhood and, as a consequence of that association, she gave me free drinks or made sure I got doubles for the price of a single. I still thought she was hot, but in conversation she explained to me how she was living with a great guy who had a great job as an auto mechanic. As much as my 21-year-old lust urged me on, I fully realized she was no longer available and that I should focus my attention on the dance floor and all the hippie wannabe chicks looking for good weed and getting laid.

On my last night there, Darla continued to feed me free drinks but was not aware I was leaving for Vietnam and that most likely it was my last evening ever at the Hollywood Palace. As the evening progressed to the point of me being half-past falling on my face, I came to the bar for a last look at Darla. I wanted to tell her goodnight and so long; and, over the thunderstorm of club noise, I said to her that I'd see her in a year or so and told her where I was going, Vietnam. Her attention was immediate and she looked very carefully at me as though she was seeing something beyond me, something I didn't understand. Her intense look was not long but was deep enough to allow the formation of a universe to go unnoticed. She touched my face, almost a stroke, and then slowly pulled me to her lips. There was no sound, no music. I knew I was drunk but even with all that alcohol in my blood I understood why my body was reacting as it was. But there was something else there that even to this day I understand, and it is the immense power that such special women have.

That evening as I continued thinking about my meeting with Dennis, the dead and gone, of a hero on Hamburger Hill and Darla, it all brought forth this huge mixing bowl of thoughts and emotions. Torn and faded images fired off in my brain and my head was full of weird pictures and smells, and I could even recall her taste even though I never saw Darla again after that night.

All of this return to yesteryear rushed back to me because of a half-assed reunion during the rains of a Baltimore monsoon. There are connections in this life that cannot be known and are hard to understand. This whole series confuses me still. How did thirty-seven years disappear? What of these folks and their stories and what was the truth of that kiss? And for the special woman that kissed me and taught me the power of a kiss, I am sorry, for your name I cannot remember and I do not know why.

CHAPTER THREE
Back In Vietnam

After years of working and raising children, I found myself wanting to go back to Vietnam, to see what the place looked like after all the years under the communist leadership.

My first trip back in 2001 included my wife, Pati, who was somewhat hesitant about going, but after a few conversations she signed up for the trip. That was in 2001. The trip did not get us back to Pleiku because the Central Highlands had been closed off to foreigners due to the religious rioting and suppression by the army in that area.

The following year, 2002, I went back and traveled with a friend, Joe, whom Pati and I met on the 2001 trip. So, the first stories you read in the following pages will either have Joe or Pati included.

After the 2002 trip, I started going back every year, sometimes several times a year working with a charity organization that I had joined. There will be no Pati or Joe in those post 2002 stories as neither ever returned. Joe was too busy selling real estate in New Jersey and Pati had her taste and had other things to do back here.

People ask me, "Why the hell do you keep going back to that shit hole?" and, for the most part I ignore that question as it reflects a negative experience or an ignorance of the country. Either way, it is not a question to be entertained. People who in one capacity or the other have left in the soil of Vietnam their blood, sweat and tears have earned the right to forget or forgive... or simply to move on. Please join me for a few minutes, sharing what I saw and experienced.

Of course there are many things to write about, but these are highlights, flashes of time now gone, and these memories are fading with age.

INVADER? Who? Me?

We stayed at a hotel in DaNang called the Bamboo Green #1. That's not to say it was the best, but that it was just 1 of 3 hotels in DaNang called Bamboo Green. All in all, it was not a bad place to stay and certainly equaled some of the places I stayed on R&R years earlier, but in no way would you call this place a Sheraton.

There are many fine stories I have to relate about the people I met in DaNang, but the one story that sticks in my mind right now relates to a young desk clerk, about 23 years old at the most, as she was a recent graduate of University.

Apparently, and from my own observations, most foreign visitors to Vietnam are German and French, usually travel in large groups, and most English speakers are back-packing Australians looking for adventure in the green and mildew of Vietnam, which also describes their general appearance. Anyway, we Americans are few and far in between and my buddy Joe and I were the first Americans this young lady clerk had met. She spoke English rather well and a conversation with her was easy, not the teeth pulling exercise that one deals with when you know the person has no idea what you're saying but will insist on "yes, yes" to everything. In the conversation she asked if we had been in Vietnam before and we affirmed that, yes, we had back in 1968 and 1969. She got quiet, then asked if we had been "soldiers," which we affirmed and she was visibly taken aback and then she said, "Oh, invaders."

Well, I didn't really know what to say to that as I had never really thought of myself as an "Invader," but I figured who the hell was I to argue with this young lady? To her I was the living (older version) of the American Invader who came to their country to persecute Uncle Ho and his boys and to support the puppet government that was centered in Saigon. Well, if that's what they're teaching the kids, so be it. It's their country and if they want to call me an invader, well I don't really care. I just thought to myself, "don't mean nuthin," and told the young lady that was a long time ago. However, the story doesn't end here.

That day while in Hoi An with our guide, an ex-interpreter for the USMC who called himself "Tango," Joe and I related the story of "Invader" to Tango and how amused we were by the whole affair. Well, Mr. Tango took it upon himself to report the story to the Hotel management, who in turn dressed-down the young lady for repeating what she learned in school. It appeared to me that is was alright to call us invaders as long as we were not in their country and spending dollars $$$$$. "Money talks and Bullshit walks?" I wondered.

Mr. Tango told us the following day that he had reported the young lady for her transgressions to management and was subsequently surprised when Joe and I both gave him some shit for breaking balls on this kid.

That afternoon I found the young lady behind the desk and being as helpful as possible as was her want. I asked her directly if she got in trouble for referring to us as Invaders, and she affirmed that the manager had talked to her and had informed her never to use that word with Americans again. She said, "I learned a lesson." I explained that neither one of us took offense as we understood none was meant and I apologized to her for getting her in trouble. Jobs are not all that easy to find in Vietnam.

Anyway, it was a valuable lesson to me all in all. The country we knew does not exist. Our efforts have been reduced to the villainous and we are but INVADERS in the history classes of Vietnam. But, so what? We did what our country asked of us and we did it well, without shame. It was the lack of political will that lost Vietnam, not us. Today's reality is that it's their country and they can damn well teach what they want in their history classes. But in the end, their history classes don't mean shit when it comes to standing up to the power of the invader's dollar. Maybe I'll think differently of this whole episode in the future, but usually I stick with first impressions.

As a postscript to this story let me add this little piece of information. I see an acupuncturist on a regular basis for treatment. My doctor was born, raised, and trained in the People's Republic of China. She and her husband, also native Chinese, are good friends of my family.

While in Helen's office, just after returning from Vietnam, I related the story of INVADER to her as she was inserting the needles into my body. She listened closely as she practiced her craft and then took a small break when I ended the tale. She looked at me, smiled slightly, and in her matter-of-fact reply said, "Oh yes, that's what we were taught, too." Hell! My discovery was but old news to her.

Both Helen, her husband, and their two children are happy to be in America. By not being native-born they help me to understand what a unique position we, as Americans, have in this world. We have a special gift living in this country and we need to stop and smell the roses every once in a while, INVADER or not.

NEVERMORE!

Finally, we had reached our Saigon hotel after the long journey from Baltimore. The room was nice, nothing wrong with it; everything was clean, the bathroom was large and well lit, and the bed was comfortable to my flop-on-it testing. Pati was quick to find the chair in the corner and collapsed, just too pooped to pop.

Pati and I were too tired to talk at this point, but a small tapping sound was coming from the door. It could have been no more than a couple of minutes since we closed it. I tried to ignore the tapping, but it persisted enough to get me off the bed in a gruff. I looked through the fish-eye peep hole and could see only a really distorted woman on the other side of the door. As I watched, she again pecked at the door with her hand.

I made a lot of noise unlocking the door. Before me was a petite and pretty Vietnamese girl, maybe eighteen, maybe. She wore a well-fitted red dress which made her very presentable, but in the creepy sense that she may have been going to a high school dance.

She beamed what she must have known was a striking smile just after I said, "yes?" She pointed her slender finger at me and said, "You and me." It was not a question. I was a little startled and just automatically said, "Excuse me?" Seeing my surprise, she continued but this time with a pointing back and forth between me and her tiny self, "Me and you?"

What an offer, room service of the best kind! The "What would she look like naked?" question flashed through my brain. But almost at the same instant my body cringed, and boy did I suddenly feel old and creepy. If anyone was going to sleep with her it should be a young man who has her heart and not some old fleshy fart like me. "No thanks," and I closed the door.

Pati wanted to know who was at the door. I told her something to the effect that the short encounter was with a local room service prostitute. Of course Pati dismissed me and my comment. Pati has spent years of marriage listening to my exaggerations and silliness and this was just another one to her. "Come on, really, Jer, who was that?" Again I explained the whole thing to her, except for the part where I wondered what she would look like naked. "I don't believe it," was Pati's reply. "OK, but that's what it was," I hastened so as to end that train of conversation. We both needed down time and sleep.

Again, the tapping returned to the door, but this time with a little more authority, harder. "Maid" was the announcement from the other side of the door. I spied through the fish eye lens again and, true enough, it appeared to be an older lady standing there with her janitorial cart barely visible.

I opened the door and the cleaning lady came directly into the room saying in her best Vietnamese English, "turn down your bed" and proceeded directly to the bed. While the maid was pushing her way in I noticed the young woman in red standing there, her hands on the cart. She flashed another smile but stayed put as I turned back into the room closing the door on her beaming face. The maid pulled up short when she saw Pati, but within a fraction of a second was back to her business of turning down the bed. I knew her real reason for barging in was to sell the young woman to me for a few hours, but that was brought to a screeching halt with Pati sitting in the corner chair.

The maid was quick about her business and left without another word or a sideways glance. But as soon as the door was closed a terrible squawking came from the hallway. Of course I have no idea as to the cause or the effect of the Vietnamese verbiage, but I do know the door was unpecked the rest of the evening, and for that matter the remainder of our stay.

I turned to Pati with the maid's departure, "Hey Pati, guess who was out there with the maid?" Without giving her a chance to answer and quick enough to catch her before she could nod off I continued, "the little girl in red." Perplexed, but still quick enough to catch the connection, Pati replied, "Do you think that old lady was selling the girl?"

"You betcha! Do you want to shower first?"

And the Beat Goes On...

While in DaNang our guide, Tango, decided it might be a good idea to visit the factory and warehouse district. God only knows why. But I guess it was a slow day and my buddy Joe was up for the drive because he knew the area from his time in and out of DaNang during his tour of duty.

There were still scores of US military warehouses in use in the area, as well as many recently constructed factory buildings. As we motored along the new four lane road that lead to the beach, I noticed off to the left side a large group of what appeared to be factory workers. They were all sitting outside the factory in various size groups. About these groups were other men and women standing around as though they were talking or instructing the people in those groups. It made me curious. Was this break time, ala Vietnam?

Our little trip through the factory district ended with the road terminating at the beach. It was a wide and beautiful beach with no one present, which made it very inviting. There were no Viets, no dogs (of course) and no tourists oiled-up and frying in the sun. I knew that October was just the start of the tourist season in Vietnam and Tango confirmed that soon enough the beaches would be covered with French and Germans tourists escaping the European snows and cold.

Tango invited us to walk on the beach, but both Joe and I declined. I can't speak for Joe, but for me it doesn't do a whole lot for me to walk on the beach, except at home with family and loved ones. So while it was a pretty day and the surf was certainly inviting, I didn't want to get sand in my shoes for nothing.

We returned the same way we came, but this time we were a whole lot closer to the factory and the groups of people. It was becoming clear to me what was going on. As we passed, I noted one particular group of individuals where each person was cradling an M16 assault rifle and all were belly crawling under hastily-constructed barbed wire entanglements set up on the factory grounds. The people seemed to be going through the numbers, not really getting into it, but doing what was needed so as not to get thumped by an instructor for fucking up.

I asked Tango to explain what was going on with this M16 and barbed wire-fest. It surely seemed to me to be an odd way to spend one's morning break. Personally, I like coffee and the internet, but these people, men and women, were belly crawling under frigging barbed wire. Anyway, Tango told me that once a year the government came to the factories and provided all the workers a day of military-style training. All workers were invited to attend the lectures and hands-on training, and I'm sure it was an invitation that no one turned down. Interesting, I wondered what they did on rainy days.

I'm don't know what the idea is behind the central government's mass military training of workers, but I have a feeling it has more to do with control of the populace rather than fending off "U.S. Invaders" hitting the DaNang beaches as the Marines did in 1965. For a second I thought the trained and armed workers would be used to combat the next group of French tourists charging off a cruise ship, forcing them back to the ship to use deodorant before coming ashore.

Seriously though, it was sobering sight. Vietnam is an armed camp. Its people are controlled through multiple layers of military and police, stooges who rat on their neighbors, and an education system that openly discriminates against those persons coming from families with "bad biographies." Meaning that if someone in your family associated with or helped the American Invaders or the "puppet government in Saigon," your whole family was put on the shit-list for at least three generations. You are literally a third class citizen who is not worthy of reasonable consideration from all communists' administrative levels. You and your family are on the bottom of every list from the Big Red in Hanoi all the way down to the local, know-all, and tell-all alleyway cadre. From what I read in the Vietnam newspapers and on many internet sites, it appears the Hanoi government wants to open up their controlled economy to grab as many tourists' dollars as possible. But in reality as I observed it, it appears the government's need to control the populace, especially in the old south, is more than their want to open up and that their approach is Stalinist all the way. Even though we Americans and the Republic of Vietnam are long gone, the war still drags on in communist Vietnam twenty-seven years after the fall of Saigon.

The beat goes on, the beat goes on,
Drums keep pounding a rhythm to the brain,
La de da de de, la de da de da

In Kontum, Vietnam, 2007

"My Family's Ticket Out"

It was Saturday, the twelfth day of October 2002. Joe and I were waiting around outside our hotel in DaNang, Vietnam, for our ride to the airport. We had spent three nights in DaNang and this morning was to be the final paragraph in a sad story. A lady in her late forties approached Joe. She wasn't a stranger and she was not intimidating even though her attire included a bush-style hat with the big red commie star squarely on the front above the wide brim.

The story had started the night before as we were out on the town, enjoying an evening of cyclo touring and shopping, or haggling for the hell of it, when a lady approached Joe. I was engaged with a saleslady so I was off-limits but that made Joe fair game for the next person with something to sell. Joe was basically just hanging out waiting for me and enjoying a cheap Marlboro (fifty cents a pack), and being the gregarious person he is, he allowed the lady to engage him in conversation. She seemed harmless enough as she didn't have any wares for sale, and she was a bit long in the tooth to be serious about prostitution. But, as it turned out she had something a little more shocking for sale.

The lady spoke English well and was accompanied by her son, a fellow in his late twenties I'd guess. When she saw us she figured we were Americans. And she also correctly figured we were ex-GI's. She was two for two and this gave her the strength to further the conversation with Joe. After several minutes of being polite she quickly got to the point that she was looking for someone to marry her 22-year-old daughter. And, as far as she was concerned, it would be really great if that marrying someone was an old American GI. That way her new husband would be a man of means, with sufficient funds to marry and ship his new bride back to the United States. Following, the lady acknowledged that if her daughter made it to America, the rest of the family would be eligible to join the daughter in the land of the big shopping malls.

Of course the only real big flaw in her plan was the fact that both Joe and I are married – long time married. Even though the photograph of the young lady presented a rather good-looking and westernized person, it was to no avail.

"Sorry, we're both married; no way. It's impossible. Ho Chi Minh will shit gold before it happens."

None of the explanations sunk in, and the little consequence of us being married, well, we could figure out a way around that. When I joined in the active conversation the lady offered to sell her daughter for one dollar if either one of us would arrange to take her to America. I really don't think this lady had a clue what it would take to get her daughter out, but we could tell she was desperate, and that was the part that hurt. Here was a family, desperate to leave a life and a country that they had no desire to continue being a part of, and their only item to bargain with was the body of the daughter. Anything was the plan, anything to get out.

We were invited back to the house to meet the daughter, but the cyclo drivers were nervous about that proposal. Tom, my driver, kept telling me, "You don't know Vietnam. I live here. This could be trouble. Maybe you out on the street, that's ok, maybe you in the coffee shops, that's ok, but you go to her house, could be big trouble. Maybe you no trouble, but me...yes, I think."

Being the free thinkers we Americans tend to be, Joe and I said the hell with the admonitions not to go and we went anyway. We thought we could be kind enough to this lady to meet her daughter. And of course there was the curiosity of it all, as one who slows down to gawk at a traffic accident on the interstate. The cyclo drivers took us, but boy were they nervous... and since they were working they didn't want to ditch their fares, us.

We made our way down a well-paved alleyway to the end house and entered. There was a small buzz going on in the alleyway because of the two foreigners who had come there. People were looking in the window as we seated ourselves for the meeting. Privacy, it doesn't exist in Vietnam.

However, the meeting was not to be. The next thing we knew our day guide Tango showed up. That was a shocker. Where did he come from? Was he following us? As it turned out, he lived two doors up and had come to investigate the buzz in the alley, as did many others who lived in the neat alleyway. When he spotted us, to his surprise, in the neighbor's house, he came in without hesitation and dismissed us from the property. "You cannot do this" he stated. "You must leave." Of course, being Americans from Baltimore and

Philadelphia, our first reaction was, "What the fuck?" Maybe all this was just a precursor of things to come for Joe and me in Pleiku, but at the time it was no big deal.

Tango, our DaNang guide, who was once an interpreter for the U.S. Marines during the war, suffered after the war, and that was the basis of his fear. In 1975 he was captured by the North Vietnamese Army, and he told us that he had accepted his death when an angry NVA Liberator pressed the muzzle of an AK-47 against his head. For whatever the reason, Tango was not executed that day but was sent to one of the infamous "re-education camps'" where his indoctrination into the enlightenment of the communist way began and he was given sight to see the error of his ways with the Americans and the Saigon puppet government. He was just glad to be alive, reeducated or not. Tango was released from the reeducation camp after six years but was forbidden, under the threat of death, to speak a word of English. And that is the way it was for fourteen years. With the opening of Vietnam to tourism, Uncle Ho's men needed people with English skills. Many of the tourists, with their free flowing dollars, would be rolling off of the tourist's ships and needing interpreters, tour guides and other English speaking accommodations. The State needed him, and Tango said it was nice to work again.

Joe and I stood to face our guide and started to protest. But we both could see the desperation on the faces of the guide and cyclo drivers. We looked at one another and quickly concluded that it was the right thing to do, to go even though we really didn't understand why, and so we did. With the sounds of continuous pleadings coming from Tango we mounted the cyclos and the drivers beat a retreat up the alleyway and headed quickly into the streaming motorbike- infested road. I don't know about Joe, but I felt like we had just robbed a 7-11 or done something else criminal and we were in a desperate escape, but from what? "SHIT!"

The cyclo drivers huffed and puffed a little to get our heavy American butts up the alleyway and out into the street and we turned down several streets to take a cruise along the river front. Joe and I didn't have much to say after the ruckus, and the cyclo drivers were keeping a little quiet, too. In time, we came back to the little street cafe across from our hotel and we all sat down to have a mixture of drinks and smokes. The cyclo drivers were still worried and in the following minutes they became more so.

Our guide, who had chased us out of the alley, appeared at the cafe. He had reported the "incident" to the local police station and came by to tell us about it. Tango said the "lady" had gotten quite upset when he told us to leave and she in turn caused quite a ruckus when we peddled out of the alley. Tango didn't like her cussing at him and he reported the whole thing to the DaNang 5-0. For sure the cyclo drivers saw their jobs going into the shitter. These guys were worried. They were not trying to get us to sleep with this or that person, go to a bar, and stop by a shop or whatever... they just worried about the police. They were honestly afraid. I didn't understand that type of fear.

I suspect that Tango was doing what he thought was necessary to protect his rice bowl and was not about to play "I've Got a Secret" with the DaNang police. So he told them first. But a funny thing happened. Nothing, not one soul came by to investigate. Maybe they were too busy at DaNang's equivalent of a donut shop, who knows?

The evening wore on and I decided it was time to retreat to the safety (?) of the hotel room and get a couple of hours of sleep before leaving the next morning for Pleiku. At the time I was looking forward to the trip into the Central Highlands and the trouble in DaNang seemed... well, not worth it. Good night!

The next morning was good, nice and clear. Breakfast was good and the packing went smoothly. We checked out with our toothy, grinning, socialist desk clerk and made our way outside to await our ride to the airport. While waiting, who showed up to say good bye? Yes, it was momma, daughter, brother and a friend. Additional photos were handed over and the mother insisted that Joe pass on the photograph in America to find a husband for her daughter. She went on and on about her daughter. As Joe engaged the family in conversation, he had some small exchanges with the daughter. Joe asked the attractive young lady if she didn't want to meet and marry a man her own age, someone who would love her. Her reply was as immediate as it could be following the last syllable of Joe's last word falling off his Philadelphia inspired vocal cords, "I am my family's ticket out of Vietnam." The statement pushed back Joe's ordinarily free flowing conversation and slowed the meeting as the gravity of the words took effect.

I guess, in retrospect, I could wonder if this young lady had been raised, at least since puberty, with the idea that her sex would be a possible passport not only for her but for her family as well. Or had this been a realization that she had come to after possible conversations with friends, families and other associates? I'm sure this young lady would make some westerner a good wife, but it would need to be a man willing to bring over the family, too, and that is time and money. The pleadings didn't end until we left for the airport. It was pathetic and sad. It was Vietnam 2002.

At the DaNang airport Tango bid us farewell. In the slack time before our departure he told us how much he loved to play tennis. In fact, he talked at length about this simple joy in his life, and as we were departing, he casually asked us to send him a new tennis racket. He wanted a Wilson tennis racket, and it would be appreciated if we would send him one as soon as possible. He even gave us the racket model number. I forgot it even before I boarded our flight for Pleiku. Damn, too bad, Tango.

Without A Dream

It was my first day of visiting the orphanage and all had gone well. As a last item for the day I left the grounds of the church and orphanage and crossed the street to visit the convent associated with the church. In a small room there a little store was established where the "ethnic minorities" of Kon Tum prepared and sold the type of items that had been used by the Montagnard people for hundreds of years.

After I had completed the visit, I was crossing the street again and had just reached the church side of the road when a somewhat grizzled man came up on his bicycle which he immediately stopped at my crossing point. His clothing was shabby even by the standards of the poor in Kon Tum, Vietnam, and the bike was of a vintage that had to be approaching total collapse but was somehow maintained in serviceable condition.

The old fellow spoke to me in a very clear voice, "May I speak with you for a moment?" I was somewhat taken back by the lucidity of his English and I bade the fellow to approach as I turned toward him, "Sure. How can I help you?"

The very dark Degar man was just over five feet in stature, Vietnam-thin in khaki and dirt shirt and shorts, and on his feet wore something that at one time must have been shoes. He did not introduce himself by name but proceeded to ask me directly, "Where are you from?" With an easy smile I answered, "America." I watched his face turn to a smile.

He continued, "During the war I was an informant for the American Army. Where do you live in America?" I told him that I lived on the east coast near Washington, D.C., to which he smiled and said, "Do you know Riley in North Carolina?"

How could I answer this question? It appeared to me that he was asking if I knew of someone who was important enough to him to remember his name for thirty plus years. I thought and answered as best I could, "No."

About this time my Vietnamese guide and escort, Van, came over to my side and the diminutive bike rider asked me if she was from Saigon, but before I could reply, Van answered his question, "No, Pleiku."

The old "informer," as he had proudly described himself, looked at her for a very long second, then back to me. Time had just exploded. His face lost the gleam and joy of his first words, his smile retreated to his mental jungle, and he stated quickly, as though officially reporting a point of law, "I am ethnic minority and we are not allowed to talk to Americans." With that statement he quickly retreated to his bike, mounted without the loss of a second and peddled down the red dirt-covered street without another word or glance. He was a beaten man, alone.

Nothing further was said by me or the guide. I understood his fear and the possibility of punishment for breaking the "no American" prohibition. I fancied that I understood his pain of abandonment, his longing to meet someone who shared a memory of "Riley," in a time of warfare that the present Vietnamese government calls the American War.

I suspect this small man will continue to ride his bike for more years through and around Kon Tum, thinking of Riley, thinking of promises not kept and finally, one day, putting aside that thought of yesterday and dreaming no more.

Vietnam's Ethnic Minorities: On the Reservation

In 1869 General Philip Sheridan, a Civil War Union hero, was in command of the Indian Territories which were located in present day Oklahoma. The Indian reservations were included in his command, the Department of the Missouri, and Sheridan's charge was, in part, to ensure the Indians would remain on their reservations.

While at Ft. Cobb, General Sheridan held a meeting with fifty Indian chiefs. During that time, a Comanche chief was introduced to the general. It is reported that the exchange between the general and the chief started with: "Me Toch-a-way, me good Indian." Sheridan's response was, "The only good Indians I ever saw were dead."

This crass statement has since morphed to the more familiar, "The only good Indian is a dead Indian," but what does this have to do with Vietnam?

In the Central Highlands of Viet Nam, in the area that used to be in the Republic of Vietnam prior to April 1975, lives a group of forty indigenous tribes who are known collectively as Degar or Montagnard. These peoples were native to Vietnam, but, through centuries of ethnic pressure and warfare, were finally pushed into the mountains of the Central Highlands. There they enjoyed an insulation from the lowland populations and became quite adept at life in their new lands. The mountain tribes developed their own distinct art, architecture, music and dance and subsisted fairly insulated by hunting, fishing and practicing some limited agriculture.

Their lives remained basically insular until the French established a military post at Ban Don in 1899 and introduced the first Vietnamese into the highlands, the servants that they brought along. From there it was downhill. In 1946 the French returned to Vietnam after the Japanese occupation was squashed with the allied victory in the Pacific War of 1941–1945 and established a federal government for the Central Highlands Degar. This was to establish the Highlands as a separate government from the government of

Vietnam at that time. After the jurisdiction was turned over to the last king of Viet Nam, Bao Dai, the independence of the Central Highlands was short-lived. The new government established for South Vietnam in 1954 found it expedient to relocate thousands of ethnic Vietnamese into the Central Highlands and dismissed the authority of the Degar peoples to run their own affairs.

More troubles befell the Degar with persecutions by the South Vietnam government and the Central Highlands being turned into a battleground as the forces under the communist north began to infiltrate. It is estimated that over 200,000 Degar died during the war in Vietnam. When the Saigon government collapsed in April 1975, the Degar continued to fight the communists and die until 1992 when the last members of their resistance group surrendered to the United Nations in Cambodia.

However, the government in Hanoi has been relentless in their continued campaign against the Degar peoples, a campaign steeped in blood and terror and ending with the nearly decimated population of survivors being rounded up and placed into non-nomadic villages and rendered into citizens of the Socialist Republic of Vietnam, second class citizens on the reservation.

In modern day Vietnam, the term "Montagnard" is not politically correct and is not recognized by any of the good citizens. These once-upon-a-time Montagnard are now… the "Ethnic Minorities." Do not call them anything else around a good citizen of the Socialist Republic as they will either ignore you or correct you right away.

There are international organizations involved in protecting the Degar from persecutions, and these folks are fighting the good fight against the government of Vietnam and their roughshod treatment of the mountain peoples. There is good reason for these organizations to exist, but, in the end, the Montagnard will be on the reservation just as the American Indian was.

In a romantic sense I feel sorry for the Montagnard and their loss of a way of life that has been theirs for centuries, but the hard cold fact is that they are losers in the nation building of Vietnam and they will become bigger losers the more they reject the government that wields singular power over them. The Degar do not have the population numbers or political clout to effect change; they are and will continue to be, for better or worse, either victim or second class citizen – reservation bound. The Central Highlands are now under martial law and the military presence in the highlands is much more prominent than in the rest of the country. Resistance is met with overwhelming force.

I am not defending the government of Vietnam and their treatment of the Montagnard, but I do see a parallel to what that government is doing with the Degar and what the U.S Government did with the American Indian. I truly feel sorry for the oppressed Montagnard, I feel bad about their loss of basic human rights, their way of life and their lands, but they are people living in a country that is not my country. Of course there are Degar peoples in the United States, refugees from slaughter and ethnic and religious persecution, but being persons in exile in America no more saves and preserves their way of life than anything we can effect in Vietnam. It would be the same as removing the American Indian to Nepal to preserve their way of life.

As concerned outsiders, we need to be careful and watchful. We need to learn from our own history the cruelty that can be visited upon a nomadic, non-technical people and we need to keep the international effort alive that calls attention to any and all related problems. We need to be fully aware of the policies of Vietnam and how they are implemented, and we need to ensure there will be no equivalent to a Wounded Knee in Vietnam.

Small Talk

Small talk is defined simply as casual or trivial conversation, a way of talking with another person or persons when the conversation does not need to be of a serious nature.

In my several trips back to Vietnam in the past few years, small talk has been the almost exclusive basis for conversation with the local people. As a visitor to their country, there is no need for me to discuss history or politics, for it is a given that, for the most part, I will be seeing history and politics from a western slant all my own. Needless to say, a good argument could easily be started but not so easily ended. So, small talk is the way to go.

I was in the town of Kon Tum in the Central Highlands and found myself in a small talk conversation with a man who spoke English very well. I was a little surprised by his easy use of the English language and his clear enunciations of what can be a problematic spoken language for certain Asian cultures.

We were going tit-for-tat in our small talk when he told me that during the war he had been an interpreter for the U.S. Army and that is where he had honed his skills. He enjoyed speaking English and tried to speak with every foreigner that came through town and who seemed as though they spoke the language of his former employer. The man was unemployed and lived by working odd jobs and on hand-outs because he was a marked man, having previously worked for the U.S. Army.

He proceeded with our small talk that had just gotten serious with his acknowledgement of prior employment, when he continued that he wanted to come to the United States but had been turned down. The reason for his denial was that he spent only two years in a re-education camp and not three or more. This was not small talk.

When the orderly Departure program was developed between the U.S. and the Socialist Republic of Vietnam, one of the requirements of the application process was that the person needed to have served three or more years in a reeducation camp. Therefore, this intelligent man, loyal employee of the U.S., was denied and made a prisoner of circumstances in his own country of birth.

Later, the program was modified and the requirement for a tour in the reeducation system was lowered to one year, but the "gotcha clause" attached to the new stipulations was that if the applicant was previously denied there was no reason or basis for application under the new process. Damn small talk.

Small Talk II

I was in an open storefront building, seated at a table, and in the process of enjoying some Vietnamese green tea. The tea was something to keep my hands busy and my mind entertained as I waited for a car to arrive at the taxi stand to whisk me off to another destination.

Next to me, very close, sat a man who was too young to have been involved in the war that ended in 1975, but I noticed a curiosity about his right hand. I have a habit of seeing things that I don't necessarily want to see, as I do not want to be placed in the position of asking what may seem to be rude questions just to satisfy my own curiosity. I noticed a ring on the right ring finger that looked like an American high school ring, and, as I focused I could see that the silver ring with a dark stone was someone's graduation ring of Murray County High School.

The man saw me looking at the ring and proceeded to lift his hand so I could inspect the ring closer. Yep, it was Murray County. Now this was small talk of the no language variety.

I looked at him and he at me with a broad smile when he continued our non-conversation small talk by simply stating "VC." I knew this man was trying to tell me that the ring had been liberated from its original owner by the Viet Cong. Well, for all I knew, Mr. Murray County graduate 1965 could have abandoned his ring in a whorehouse in the old Pleiku town and that the ring had circulated from then until this time. VC? Well, maybe yes, maybe corner junk shop. This small talk was starting to get on my nerves. I thought about offering this man some cold US cash for the ring. I thought that maybe I could track down the origins of this ring and find the owner who lost it so long ago, or turn it over to a grateful family of a long lost son or husband. Or I could have turned it over to a beer-gutted vet, happy with his pension and rewarding me for the effort by telling me how he had lost the ring one night while doing the dirty in one of Pleiku's raunchy brothels.

I was happy when the cab appeared. I smiled and left. No more small talk today, just let me ride in silence.

Us or Them?

Mom was seventy-two years old in 2006, and although that is not all that old anymore in the U.S., her age showed heavily on her in Vietnam time.

Mom was not an American, she did not speak a word of English and her home, the home she had lived in since 1941, was on a small back alley off a main road in Pleiku. When the house was built by her husband, Pleiku had only three roads and the population of the town was very, very small. In the jungle, which was all about, tigers roamed and elephants were not too much deeper into the darkness under a triple canopy of green.

It was Tet, New Years in Vietnam, and I and another ex-GI had been invited to the lady's house by her daughter who was also acting as tour guide and escort for my buddy, Marc, and me as we went about our business in Pleiku and Kon Tum. Tet is a fine time in Vietnam, especially now that we are no longer shooting at one another and I'm old enough and have lived long enough to appreciate the cultural specialty of the occasion.

We were sitting in a very small room that was opened to the outside by one small window and a diminutive door. There was a table with several child-sized chairs and a corner shelf with a small television upon it in the far corner of the room, but still near enough for me to almost touch. An elfish recliner of sorts provided the viewer with a seat to enjoy the programming allowed to the subjects of the Socialist Republic of Vietnam by its current government, the Communist Party. It seemed that Marc and I took up a good part of the room, and with our guide, her son, and a visitor or two popping in and out, the room was filled to standing room only.

The table was a spread with a nice selection of goodies, and we picked at the fruit and sweets as we all talked with our guide who acted as interpreter and cultural go-between. The easy talk went in different directions without aim or argument until sparked, I suppose, by Marc's retirement to the alleyway to take a smoke, the elderly mom decided to tell us a little story about the war years and cigarettes.

When the war was in full swing, many a GI found solace in having a smoke of wacky weed, marijuana. And, as it so happened, the local Vietnamese had an endless supply of the delightful herb that they ingeniously packaged in American cigarette packs. The industrious Vietnamese entrepreneur would very carefully open a pack of, for example, Salem's, from the bottom and carefully remove the cigarettes. As mom told the story, the tobacco was removed from each cigarette, replaced with well-shredded and tightly- packed "Pleiku Joy," and finally all twenty cigarettes were repackaged. When complete, the pack of Salem cigarettes appeared to be just purchased from the Base Exchange. But this was good shit, not tobacco. And the price was right…as I remember.

Mom seemed to delight in the story telling as it was her contribution to the war effort I suppose. But I didn't have the heart to ask her who got the contribution… us or them?

CHAPTER FOUR
Sharing Thoughts: Aging and Death

Why in hell would anyone want to have a couple of stories under a chapter with the name, Sharing Thoughts: Aging and Death? Damn morbid I'm sure will be said, and maybe you're correct. But I didn't want to "pretty it up" and, of course, aging and death are just natural parts of this life we are born into. Can't escape it, right?

The last story here is put together from a series of emails I received from a fellow Vietnam vet whose name is Ron. He, too, returned to Vietnam many times after the doors were open to tourism and in one of his trips he meets a school teacher... we shall call her Nguyen.

I asked Ron for his permission to print this story, and he was happy that I wanted to do so since it would be an honor to have their story remembered.

So, both tales refer to our inescapable human mortality, but I do not consider either to be morose or morbid... just life, just the way it was and is for Vietnam: Without A Dream.

Night Stroll

The evening was pleasant, warm, and slightly breezy with the promise of rain, and not too much traffic with the hour getting late and tomorrow just a few hours away for the workers. The tree-lined street held jumbles of motorbikes parked on the sidewalk, Hondas and the nondescript Chinese rice burners. There were some late night diners, shop owners now eating dinner and assorted others about who live in or on the city of Saigon.

I had just finished dinner and was walking back to my hotel about three blocks distant hoping to beat the rain that sent sprinkles. As I walked, several propositions came my way for massage, girls, and the last offer from a greasy pimp proffered a massage with "boom-boom." Hell, apart from stupid Viet Nam War movies, I had not heard that term since I left Viet Nam decades ago in 1970. "Boom-boom" I thought… hahahahahah. My reply, with a smile, was "no thanks" and a quick wave off with my right hand as I continued to walk in the direction of my hotel. Dancing along the curb the pimp continued his sales pitch, " Give you Goooooood massage, she take care of you, suck your dick, boom-boom." Again, "No thanks," I replied.

Mr. Pimp was somewhat irritated, I guess, still dancing along the curb, off and on into the street. I wasn't buying what he was selling so I guess he decided to take one last shot… not at selling, but at insulting. He stopped walking, stood in the street and shouted, "You too old, huh?"

I quickly thought of telling him "yes" too old, and I was so old I probably fucked his mother back in the 60's… but I decided not to go to war. And besides, I am sure that in his line of work he has heard it all anyway, so why waste my breath. "You too old," he repeated. I decided to end the conversation with a polite "fuck you" and then crossed the street to the block where my hotel offered respite.

The neighborhood dogs were out for their evening constitution, the street pho restaurant was alive with night workers grabbing a hot meal over on the corner, and a fresh breeze sprang up from the river just two blocks away across several trash-strewn empty lots. The grocer was open and the midnight bread man was heating a small loaf of banh mi over a charcoal fire right next to the steps to the hotel. I stopped at the foot of the steps and looked up to the second floor where they ended with the small lobby door. Too old? Yeah, maybe he had a point after all.

And All I Can Do is Think of Her

I woke up and I was sixty. Where had the time gone? So fast, so very fast that I could not believe I was what I once thought to be ancient. I don't feel in my heart and soul that I am very different from fifty, forty or even thirty, but I know in reality I must be.

Even though I am sixty, I also woke up with the thoughts of her. I could hear her, I could smell her gentle body perfume, and I could still taste her. But how could this be? I'm sixty and I am not supposed to feel this way... I am not supposed to think this way but rather I am supposed to be totally different than what I feel inside... and I oddly think it is the monster of my own creation. I look at myself in the mirror and I see the road map that time has etched on my face; I see and feel the extra pounds that I've allowed, through slothfulness, to collect on my once-slender frame. It was easy to ignore me while being so many other things, while I was being all those other people I was and am supposed to be.

Time has slipped by and I realize that in so many ways that I am blessed time and time again and that I have been able to do many different things in my life, but is it because I am just lucky or is there Divine intervention? I thought I knew, but I realize that I don't.

But what of her, at age sixty, what of her and of her part in my life? I smell her, I taste her and I want her... but what? I am not sure about tomorrow, and right now, well, I don't care about tomorrow too much. Prostate Cancer has brought me to the understanding that my time is finite... and that time, whatever it may be is now shorter than it was the few minutes ago that I started writing this down.

I hear the clock ticking just as clearly as when I was a thirteen-year-old boy sitting at the kitchen table with my maternal grandparents. It was January and my father was attending his mother's funeral, a woman I did not like much. There was snow on the ground and it was cold, very cold. The old kitchen was warmed by the wood-burning stove and my grandparents, lost as what to say to me, supped coffee and listened to the clock tick away. Maybe they knew their time, too, was limited by that clock... tick, tick, tick… every second gone, never to be regained. The clock and quiet bothered me then and it still does today as I realize the lesson about the shortness of time that I was learning from that clock at age thirteen; and today I still hear it and I still see my aged grandparents... supping and passing the seconds in a small place, surrounded by the cold of tomorrow, waiting for eternity.

So I woke up this morning and I was sixty... and all I can do is think of her.

A Letter from Lynne

NOTE: I met this lady, Lynne, her sister and mother at a Vietnamese shopping center in Northern Virginia. In this short email letter, Lynne goes from a tragic war, destruction of family, and then the renewal offered and found in the United States.

Hi Jerry,

My brother's name Phero Tran Cao Duc, his Catholic Saint's was "Phero" means Peter in English.... only 23 years old.

He came back on April 17th, 1975, (to Vietnam) somebody shot him on April 19th. * That was sad, my mother decided to stay back with him after the funeral. We left the country by the end of April with my older brother and his wife (who has only one leg), this brother got shot in 1968.

We got to Guam, then to CA, then to VA. From there, my father arrived with my other older sister with her husband.

To make the story short, I came to the states when I was 15 years old, my father passed away when I was 19, got a scholarship with The Catholic University of America, graduated BSEE in 1983, worked for Ford Aerospace as a system engineer, went to CA in 1986 (got married), my hubby was my classmate at CUA, he then got a scholarship at Stanford and landed a job in CA. That is why I am out here on the West Coast.

Next time, if I have time to talk....I'll tell you my journey to America, it is quite amazing.

Lynne

* Duc Tran had completed his flight training in the U.S. in March 1975. He was advised to remain in the U.S. because of the imminent fall of the Saigon Government. Duc refused to abandon his family and his oath. He

returned, came to his assigned squadron in Saigon, and two days later was shot and killed by others in his squadron who were defecting to the invading North Vietnamese Army.

Lynne and her family are doing well, very successful, and Americans all.

Letters from Sai Gon #4

Night Noise

There's a banging outside
As the last drops of steaming rain
Fall eight stories below
And I want you, your face I see.

I open the window and kill the air
As I crave the smell of the city and the rain
And to listen to the night noise.
I breathed deep of the Sai Gon Street below
And I flashed back a life time but that was never you.

My soul wandered back there to 1968
And I searched for a meaning to it all
But there was none I could explain
As it all had brought me here…. to you.

Briefly a silence of noise and then the rain again,
And I feel you here with me.
Shivering, I remember your woman's moisture,
and I want you more.

Nguyen's Tale

The following story is presented in the form of letters written by a young woman, Nguyen, in and mailed from Vietnam and commentary offered on those letters by the American recipient, Ron. The letters document a unique story from July 1997 through October 1999.

It is a bittersweet love story that takes place in Vietnam several decades after the war was over. The two people are an American man, an Army Veteran of the war, and a young Vietnamese woman with no memory of that war. Ron was perhaps the first American she had ever met. Nguyen was trying to earn a living as an English teacher for small children and the street vendors in the coastal tourist town of Nha Trang, where Ron and she first met. At the time, Nguyen was twenty six and Ron, fifty one.

Acting as the editor, I will invade very little of their space and only do so in an effort to clarify an item that may be of interest. Additionally, I will not edit the letters but present them as they were presented to me. The English is a little confusing at times, but as an interested reader, it is necessary for you to read the words of the young woman, Nguyen, as she wrote them. Spelling has been altered where needed and just a few sentences were changed in the commentary written by Ron.

INTRODUCTION

By way of introduction, it will suffice to say that in 1997 Ron was returning to Vietnam, as were many of us, to see the land he had learned to love even during the time of war. It is too difficult and beyond my powers of expression to tell you how it was to "come of age" in Vietnam, during a war and a time that was, and remains, very alien to the American experience.

We went back trying to reconnect our memories with today's reality, and, I suspect, that many men have experienced a rebirth of purpose by looking so deeply into their past and evaluating their experiences as a kid, (and, yes, we were but kids) by the understanding that comes with age and experience.

Many of us connected with the past by visiting the Vietnam Memorial in Washington, D.C., and it is a hallowed place, but a place where we hang our heads in sorrow for the uselessness of a buddy's death, and where we, the living, descend into the bowels of the earth and return with broken hearts and tears we cannot explain to our wives or children.

But, to return to Vietnam, to smell the odors we all knew so well, to hear the language we all knew by sounds, and to see a country much as we left it; well, that was life, and we will always chose life when we are presented with the choice. So, Vietnam was a re-connection for many of us and a way for us to assert that we did what was asked, we survived, we honor those who died, but while we can, we choose life.

I cannot speak as to why Ron returned to Vietnam in 1997, but I suspect it had something to do with choosing life.

Ron's Opening Commentary:

Jerry... I will attempt to explain my relationship with Nguyen. Our conversations here, both written (emails) and verbal (conversations) will be in the way they were received. You may take liberty to use, change, or delete any of the information that you feel appropriate.

My name is Ron. I served in Pleiku Vietnam from Mar. 2, 1968, until Nov. 19, 1969, with B Co. of the 504th Military Police. My duties at that time were to open and close Rt. 19 from Pleiku to the bottom of Mang Yang Pass and to support convoys along that stretch of road.

I met Nguyen on my first return trip to Vietnam in 1997... I was getting ready to leave Nha Trang, Vietnam for my flight back to Saigon and then the States. I had my luggage with me in front of my hotel and was saying my good-byes to Op, a street vendor. She asked me if I would kindly talk to her friend, Nguyen, as she had never spoken to an American. Nguyen was sitting at a small plastic table and was in traditional Vietnamese dress. When I sat down across from her, she would not make eye contact with me, so I asked her to please tell me some things about herself. Nguyen said that she was twenty six years old and that she taught English to some children in their homes at night and also English to some of the street vendors. During the daytime she would sometimes

Nguyen

go to school. I asked her how much it cost to go to school and she said 50,000 dong* for one month, and that because of the money some months she could go and some she could not.

I told her that I was sorry that I could not speak to her longer, but I had to catch a plane to go back to my country and maybe I would see her again. I then took a 50,000 dong note and rolled it up into by palm and shook hands with her, leaving the note, and when she looked at it... She said: "No, I cannot take" and I said, "Yes, you take and maybe you can learn more English and when I return to Vietnam you can be my tour guide." She then looked at me and started crying, and it broke my heart.

Shortly after arriving home, I wrote a letter to Op and told her how much I enjoyed meeting Nguyen and to please give her my best regards.

In July of 1997 I received a letter from Nguyen:

Note: *50,000 dong equaled approximately $3.33 at that time.

Dear Ron,

I had read your letter at Op, thank you about your kind regards to me.

I'm very merry, so I write to you today.

The first of my letter, I give to you my good regards and I wish you are always healthy and you are usually lucky in your life.

Well, how are you? How is your working? Have you any plans? Example:

When you come back to Vietnam? or something.... How about your family? What are you doing now?

And, me, I'm fine, I'm training for my studying now, in my part time, I'm teaching English for some people, and I'm learning German. You know, I can speak French.

You know? I can guide to tourist but in Vietnam now, the weather is bad so there aren't any tourist. I'm very hard in my life especially about money, it has too much problem, I haven't enough money for my training, my learning and many thing which necessary in my life.

You know? The lifestyle of Vietnam isn't good, very hard, especially to students so I'm most sorrow. I'm trying to overcome difficulty. I hope that in the future my life is better than now.

You know? I never before get money from someone but my mother and I'm very shy and very emotional. I think that you are a good man, you are kind hearted, I'm very happy about meeting you, I never forget that thing when you looked me, I would see your eyes, it is very kind hearted.

You know? I couldn't say hello when you went back to your country because I'm shy. I'm very sorry about it. I hope when you read my letter you can see? I'm very sorry. It is from my little heart.

I hope you write to me soon, I very pleased. Now I put my pen. Give my regards to your family, I wish you stay in good health.

Friendly,

Nguyen

My Address is: Nguyen Nguyen Fuong

32 Lam Son

Nha Trang – Khanh Hoa Vietnam

Letter October 1997

Dear Ron,

Today I got your mail. I'm very merry and want to say thank you about that you write to me with kind words.

The first of my letter, I give my regards to you and your father. I wished you and your father is healthy. Your working is good.

How are you Ron? What are you doing now? In you country is it cold? Keep your health Ron. I looking forwards your coming Vietnam. I want to see you soon. I had given your regards to Op and she is very merry. She says thank you about that.

Now I put my pen see you in the next my letter. I write to you longer. The last of my letter, I give my regards to you and your father. I looking forwards your letter and I keep you picture in my heart.

Friendly yours,

Nguyen

My new address: 15B Hoang Hoa Tham

Nha Trang – Khanh Hoa

Vietnam

Tel. (84) (58) 72752

If you call to me, you must write to me a letter. You must say what time you want to call to me and when you want to call to me. I wait your calling.

Ron's Commentary:

In a letter to Nguyen I told her about the French/Vietnamese girl I lived with in Vietnam in '68 and how I had traveled to Pleiku in 1997 trying to locate her. This is Nguyen's reply to me.

Nov. 22nd. 1997

Dear Ron,

I had received you letter, I'm very merry and moved to write to you now. I wish you are healthy in the first of my letter. Everything is good to you and your family. Don't forget me, write to me more and more. That thing make me happy.

Dear Ron...When you write to me you had said to me about your girlfriend. I look at your eyes, I know you are sad. I think that you love her very much. I'm very moved. I think that if we are friend together, I want you are happy. Your happiness is my happiness too. If you want me to help you find your girlfriend, I will do it. Don't worry it is easy for me because I am Vietnamese. I can contact radio station and put something in Saigon paper. If you don't want to do it I'll don't do it. You said maybe she have a good family we should not bother her.

Dear Ron...Don't worry about me about money with me, money is not important too much. If I have more money, I can live with high life, If I don't have money I can go to work. Many work to earn money. Happiness is very important with me. Special is I have a good friend as you.

There aren't enough pages to express everything that I want to say with you, but in my heart you are a kind man. I think of you very much. I always remember your eyes, your smile.

Dear Ron! I looking forwards you coming back Vietnam. Every days and very time when you come back Vietnam you send me a post. I'll go to Hanoi to catch you. You can see my mother. Don't worry anythings. My mother will like you because you are a kind man.

Now I put my pen. Give my regards to your family and friends. I wish you are happy. Your father is good. Keep your health. I'll write to you every month. I looking forward to your letter.

Friendly Yours,

Nguyen

Ron's Commentary:

In March 1998 I again returned to Vietnam and took ground transportation from Saigon to Nha Trang and when I got there Nguyen was waiting for me in front of my hotel along with the vendor woman Op. Nguyen said that she had been waiting four hours and was very worried about me. We talked for about an hour and made plans to meet the next day.

Around 10 a.m. the next day Nguyen came riding up on a bicycle that she had borrowed from her neighbor. She asked that I rent a bicycle from my hotel and follow her to the market because she wanted to buy some things and fix me a meal in her home.

Her home had only one 8 x 14 ft. room with a concrete floor and holes in the walls big enough to put my head through. There was no running water or plumbing of any kind and only one small light bulb suspended from wire in the ceiling. In the room was just a small bed with a straw mat and small table with two chairs and none of this did she own. In the corner of the room was a bookshelf with maybe a dozen books. The tin roof did not completely cover the room so when it rained she had to bail water. I couldn't believe her living conditions and I said, "Nguyen, you have nothing." She said, "Look, I have books. With books I can learn and that is what is important to me."

She cooked a meal of rice, crab and fish on a little one burner cooker. When I was eating the crab, they tasted kind of gritty and when I held one up to the light bulb there were all kinds of ants crawling over it and when I told her she said, "Yes, ants like crab, too."

I asked Nguyen if she would take me around Nha Trang to some good tourist spots and she told be that she did not know of any. I said I thought you told me that you have lived here for five years so how come you do not know. She replied, "You know, I go to school in the daytime and at night I have to teach some children in their homes or teach English to vendors. I have no bicycle of my own and I have no time to go anywhere." I then told her that I wanted to take her to Da Lat and she said, "Really?" I said, "Yes, I think it would be nice for you to go on a trip." She started jumping up and down like a child and said she would talk to her teacher and her students' parents and see if she could get off for a couple of days.

The next day I met her in front of my hotel and she was real excited and said that she could go for three days and that we could make plans. We looked in my tour book and decided on a hotel where we would stay in Da Lat. I gave her money and told her to purchase two tickets on an air-conditioned mini bus. That excited her as she had never ridden in a vehicle with air conditioning. She was to tell the bus driver to pick her up at her home and then pick me up at my hotel. We would not sit together on the bus because people would talk bad about her and when we got to Da Lat we would then get separate rooms.

The next morning when I got on the bus I did not see Nguyen and so I asked the bus driver about it and he said that he went by her house but she was not there. When I got to Da Lat I checked into the motel, took a shower and then laid down and fell asleep. A few hours later there was a knock on my door and a girl from the lobby said that Nguyen wanted to talk to me and I said, "Oh, she's on phone?" "No" she said, "she is here in lobby....."

When I got downstairs Nguyen started yelling at me for not waiting for her and I told her that the bus driver said that she was not home. She said he lied because she waited for two hours and he never came and so she had to get a ticket on a Vietnamese bus with no air conditioning and ride with ducks, pigs and chickens and she was very sick. I told her I was sorry and that I thought the bus driver was a very stupid man and she said, "Yes, he must be a very stupid man."

Anyways, I got her a room and told her to take a shower and that when she felt better she could come and knock on my door. I went to my room and took a big shot of whiskey. No sooner had I set the bottle down when there was a knock on my door. It was Nguyen and I asked her what the matter was and she told me that she did not know how to work the shower because she had only washed from a pan before.

I went to her room and adjusted the water temperature where I thought it would not be too hot for her, and when I turned around, she was naked. She then got under the shower and started acting like a little child singing and dancing and then she stopped, spread her legs and took a pee. She looked at me and said, "I make water." I said, "Yea, well that's alright. When you get done, dress and come to my room and then we will rent a motorbike and go get something to eat."

The next day we booked an all day tour to a couple of Montagnard villages, some temples and a couple of waterfalls. The following next couple of days we just rode around the countryside on our motorbike and went to the Valley of Love.... After three days it was time to go so I purchased her a ticket back to Nha Trang and me a ticket for Saigon. We said our goodbyes and I gave her some money and told her to rent a different house that had running water. Nguyen then jumped up and threw her arms around me and hugged me tight.

April 15, 1998

Dear Ronald,

I received your mail and your letter including the birthday card, I'm so happy with the earrings and the music tape, thank you very much.

Darling, How are you? Who is your work? How is your father? how are your daughters? Give my regards to them and say hello to your father and best wishes to them.

I'm fine, I had a good work at a tourist office with income 600,000 a month and if I go for tourist guide I can get 3 U.S.D. a day.......I bought my own bicycle. I'm very happy now. I thank you and your father very much I never forget everything that you and your father doing for me. Now I only wish I have my own home.

I'll save money from now so maybe 5 years later I can get my own house. Darling, in the future when you retire, can you live in Vietnam? Maybe I can get a house and we can live in Vietnam. It's not so bad, do you agree with me?

Darling, what will be, will be I just want us to always be happy... My mother write to me and want to say hello to you and your father and thanks you and your father about helping me with my live and she wants to see you soon.

Darling, Best wishes to you, take care yourself and forget me not. I think of you and missing you very much.

All my love for you,

Nguyen

Ron's Commentary:

Jerry, I wrote to Nguyen and told her that my father would lend her some money so that she could buy a better job.

Note: People seeking good paying jobs in the corrupt system of Vietnam not only had to have the qualifications in both the academic and the political sense, but they also had to "purchase" their way into the job. This is what we call a bribe, but it was the norm at that time, and still is (as of 2013), in Vietnam.

April 26, 1998

Dearest Ronald,

Today is 26-4-98 I get your letter with some photo's inside. They are nice photo's but my face is dark, but I like them.*

Dearest Ronald, how are you? How is your father? How is your family?

I hope you are always healthy and everybody in your family are too. And me, I'm fine. I'm still on my school and working. I had written to my mother about your coming to Vietnam, but can't go visit her because you are very busy so she said to me that give her regards to you and your father when I write to you.

Dearest Ronald I thank your father very much he is very kind, I am very happy to accept his loan but I worry about it in the future I have money to pay back to him good, but if I cant pay back to him I'll be a ungrateful woman and I don't want any person to misunderstand me. Although my life is very hard, I don't want to be an ungrateful person...can you give me advice?

I put down my pen now. See you next letter I'll write more and more. Forget me not. Looking forward your letter.

All my love for you,

Nguyen

p.s. The money you give to me in Dalat I put in the Viet Com bank in Nha Trang so I can keep it to spend for letters. So now I have account.

Name: Nguyen Thi Nguyen

ACCOUNT: ****

I have only you so you can know about it.

Ron's Commentary:

Nguyen quit her tourist job because she had to be out in the sun all the time and it made her skin dark.

Note: Many of the women in certain Asian cultures will go out of their way not to have skin exposed to the sun, turning them a deep tan. It is considered that a person of dark skin is of low class, a person who works in the fields. Perhaps this obsession with "whiter skin" is most noticeable in Thailand where the ladies wear long gloves, long sleeves, cover their faces fully and wear sunglasses in addition to using an umbrella in the hot sun.

So here, Nguyen, although as poor as she could be, did not wish to be equated with or identified as a field worker. This is just the way it is, and judgment by others not of the culture is judgment passed without full knowledge or having walked in their shoes.

Letter of May 25, 1998

Today I get your letter again and that is my happiness. How are you darling and how is your working? How is your father? Give my regards to him. How are your daughters? They are beautiful...I think of your father like I think of my mother. I wish for him happiness and longevity.

Dearest Ron, I'm fine, I'm still looking for a work. It is very difficult. If I have about $200 U.S.D. I can get a work for all my life with average income 600 U.S.D. a year. If I have $400 to $500 USD I can get a good work with average income of 1,200 U.S.D. a year. If I don't have money I can get a work too, but I can't live with that income and today I have that work but maybe tomorrow I lose it. Do you understand? So everybody in Vietnam try to pay to get a good work. It is funny isn't it? That is Vietnam.

My friends are very happy and like my gifts which I bought in Dalat. They asked about you too much and that make me happy.

Well, I close my letter now. See you in next letter. All love for you.

Wish you are healthy

Kiss xxx

Nguyen

Ron's Commentary:

Jerry... I was getting kind of tired of this long distance relationship and with her talking about me moving to Vietnam with her, etc. so I wrote her a short note and told her that I was seeing a girl here near my home... (which I was) but I guess I should not have told her.

Letter: June 29, 1998

Dearest Ron!

Long time, I didn't hear from you. I feel sad so I write to you now.

Dear Ron! How are you? How is your work? How your father, how is your girlfriend? I wish they are good health. And me, I'm okey. I'm still look for a job. But now I work in the free time at a restaurant for friend but can make only little money.

My mother go to hospital, she very weak and I worry for her so much. I don't know why you didn't write to me for long time. Maybe I did something wrong? If I did something wrong, please forgive me.

Dear Ron, I had to change my address again because of not much money. My new address is: Nguyen Thien Thuat Nha Trang, Vietnam

Well, I put down my pen, see you next letter. I look forward your letter. Please keep yourself. Give my regards to your father and your girlfriend.

Love,

Nguyen

Ron's Commentary:

Jerry... I had never told Nguyen that I had sold my insurance business and that is why I was not working.... So she was always worrying about me not working. After about 3 yrs. of not working I finally figured out that I didn't have enough money to retire and so that's why I'm working now.

Letter dated September 11, 1998

Darling,

I received your mailing and your letter was very nice words and very nice photo's. I had sent our photo's to my mother and she wish you visit her on someday. I had promised with her about that.

Darling, I was so worry about your health in your accident recently. You should be careful and I wish you can recover from that soon. You had sold your business, didn't you? And what are you doing now? If you don't work, you shouldn't send any mailing to me. It cost much money. You should keep it for your life. My life is very hard but I'm able to it.

Vietnam was influenced by the Asian economic crisis so the life is very hard. The Vietnamese dong was loss value 1 U.S.D. = 1,400 dong and the unemployment increases so finding a good job is very hard in here. You can find a good job but only after you bribed for some person who can help you. I can teach English for some children, but I earn very little from it. It is not enough for my life so I work at restaurant as you know.

Darling, I'm fine; my mother has been sick and weak and had to go to doctor so I have to help her. I earn 400,000 a month. I give my mother 100,000 I pay for my room 150,000 & my school 50,000 and I have 100,000 left for drink and lunch, I don't eat breakfast.

Darling, I wish you are always healthy. Forget me not. Give my regards to your father, your daughters and your friends. I hope they are healthy. I send nice wind with my love from Vietnam to the American who I love who I always think of now. I miss him very much.

All my love for you,

Nguyen

Ron's Commentary:

Along with this letter she sent me a couple of jokes which I think you can appreciate.

Little boy goes to school and the math teacher tells him..... "O.K. I give you three ducks and Mr. Chong give you four ducks....How many ducks do you have?" Little boy thinks and then says: "I have eight ducks..." teacher says: "No you are wrong, you only have seven ducks." Little boy says: "No, I think you are wrong....I have one duck at home."

Two Vietnamese women talk on the street and one says: "Hello, do you have any children?" The woman says: "Yes, I have one son." "Is he a good son?" "Yes, he is a very good son." "Does he smoke a cigarette?" "No, he never smoke cigarette." "Does he drink a beer?" "No he never drink a beer." "Does he stay out late at night?" "No he always goes to bed early." "Oh, he sound like very nice son......How old is he?" "He's three months."

Ron's Commentary:

About this time Nguyen received some money from me so that she could better help her mother and also go to a internet cafe and start sending me emails and so about this time the letters started slowing down and I did not save the emails. I do have other letters that I will share but now I need to take a break. I don't know if you can use any of this shit and if not let me know now so I can stop writing and start drinking.

Ron's Commentary:

I told her I would be sending her a little money in a month or two.

Letter dated: July 30, 1998

Dearest Ron,

I received your letter at the new address and the letter at old address and the mail. The watch is very nice, I like it very much, I thank you very much it is very important with me.

Darling, why you don't go to work or you can't get a work? Or you are now healthy? You are sick aren't you? I'm very worry for you. If you don't go to work, how can you send money to me? You need for your life and your fathers life. Although my life is very hard too, but I'm able to work. I can't get a good work now so I must work in restaurant. It isn't enough but I try finding more work, but I don't know when I can get it. I try saving enough money to pay for the government for better work.

Darling, the house which I rent before they are overhauling it so I must change another room about two months then I come back when they finish it. So next letter you write to me by the address before. 15B Hoang Hoa Tham

Darling, I think of you every times. I wish you are here. I miss you so much but I'm worry that your girlfriend in U.S.A. I know about that and she is jealous, so I keep my feeling in my heart. Do you understand me? Although I love you but I don't want to break your feeling with your girl friend. What do you think about it? But I love you forever and forever.

Darling, I hope you are healthy. Forget me not. I hope your father is healthy too I wish one day he can come to Vietnam to visit my mother. My mother give her regards to you and your father.

I'm looking forward your letter. Take care yourself..
All my love for you.
Nguyen

Letter: November 5, 1998
Dearest Ron,

I had received your letter and photo's it is so nice, but one thing is more nice is the letter because it told about your returning to Vietnam. I always look forwards to that day so happy when I heard about that.

Dearest Ron, as what I said to you in the last letter, I had a plan to return back to the old place where I used to live, but when the house was repaired, I asked them about the price. They said it was more expensive. I didn't have enough money to rent it anymore so I rent another one, it is cheaper.

Dearest Ron, I dont care about the people who look at us and say bad things. I only care about what do you think about me and how is your feeling that you give to me. It really makes me happy so you shouldn't so worry about that. It isn't good for us. My mother will be very happy when she know that I have a good man, she wants to see you soon, I think so.

Dearest Ron, on 15-2 we start Tet and on 20-2 I go back to work and school. I can't wait for you come back Vietnam, the sooner the better. I'll get off work to go with you anywhere. I like it very much. I'm very proud when I go with you.

Dearest Ron, I'm me now, everything is okey I'm missing you very much and think of you every day. I want to say hello with your father and your daughters. Give my regards to them. Take care of yourself. Forget me not.

All my love,
Nguyen

Ron's Commentary:

In Feb. 1999 I flew into Nha Trang from Ho Chi Minh City and Nguyen was waiting for me outside the terminal. We took a cab a couple blocks to my hotel and we shot the shit with Op and some of the other venders. Nguyen wanted me to rent a room near where she lived, but I needed the space. I had promised her that I would take her to Hue and then to her mother's village.

After a couple days we took a train to Hue. We got a sleeper, but there were two Vietnamese guys in the same sleeper and, when we showed up they beat feet and we never saw them again. I guess they went to the hard seats? Just outside of Da Nang the train put on the brakes and damn near shook us out of the rack. We were just sitting there for quite awhile and then I told Nguyen to go find out what happened. She came back and said that the train had hit a man on a motorbike and I asked her how long we would have to sit there and she said that we would have to wait until someone could locate the relatives and then the train would have to pay them money. I asked her how much and she said: Oh I don't know, maybe 1 million dong.

About an hour later we were moving again and we finally arrive in Hue and get to our motel room. We stayed a couple days and then rent motorbikes, strap supplies onto the back and head off to her mother's village which is around 25 miles outside Hue City. Paved road, then gravel road for awhile and then we turn off onto a dirt road and down the road it's flooded so we get off our bikes and walk them through and the plugs got wet and so I had to take them out and dry them off. We no sooner got going again and Nguyen lost control of her bike and down a hill into the water she went. She hurt her arm and leg a little but nothing serious and so after drying off the plugs again we headed back down the dusty road and then onto a jungle path for a mile or so and into the village. Most of the homes were bamboo and rice straw and when we got to her mother's no one was home. An old lady finally showed up and said that all the people, including her mother were out in the fields. The village grew rice and sugar cane.

After talking to the old lady I looked around and Nguyen has her pants dropped, is squatting and she looked up at me and said, "I make water in nature." I said, " I think I'll do the same."

Nguyen told me, prior to going to her mother's village, that her mother was a Viet Cong nurse in a jungle camp outside of Hanoi and that her father was a North Vietnamese Army officer who was killed before she was born. Also, some of her uncles in the village fought for the North and some for the South, but not to worry because there is no problem now. The village also grows tobacco and when her mom shows up she is smoking a big fat rolled tobacco leaf. Mother hugs her daughter and I get the double hand shake and then all the village people start showing up to gawk, and the young children are keeping their distance because, I guess, they have never seen a white person and they are somewhat scared and when I look at them they take off and go into hiding.

I get to meet Nguyen's uncles, aunts, and 101 yr. old grandfather. I also find out that Nguyen has a small adopted sister that lives with her mother in the village and walks three miles one way to school. All the men in the village are quite interested in our motorbikes as the only thing I saw in the village to ride were cows, a water buffalo and one bicycle. Anyway, Nguyen had to get everyone together for group pictures and the guys were all hugging me and the women were all afraid of me. Later I go to hear some of the stories of the village being bombed during the war and I got to visit the graveyard, etc.

When we got back to the house Nguyen's mother had just killed a chicken for the big celebration and I watched her draw water from the well, build a stick fire, heat the water, dip the chicken, pluck and butcher it. She strung out the guts and I was told that they would be used to make a soup. Up until that point I was kind of hungry, but then all I could think about was having a good stiff drink. I had brought along a couple of bottles of hooch, and when I pulled one out one of her uncles led me off to show me his private still where he makes rice wine. We both had a drink of mine and then one of his brew (not bad).

Around dark all of the inner family showed up and the ten of us sat on the floor in a circle waiting for the women to serve the big feast: One skinny chicken for 10 people. The cooks had torn most of the meat off the chicken and put it in a bowl in front of me and the pile of bones went in the center of the group. Once everyone sat down they started grabbing for the bone pile and so I took my chopsticks and rationed out the meat into all of their bowls. The rice was fine and the (chicken gut) soup, to my surprise, was O.K.

After the feast, a couple of the guys broke out the rice wine and I broke out my bottle of spirits and we had many good cheers until one of the nephews (good size lad) wanted to arm wrestle me. I thought I could take him, but if I did then he would lose face. And I never really wanted him to beat me because then I would lose face, and so I just refused to wrestle him. He got pissed-off and wandered home. Nguyen said that I had done the right thing and her mother was happy because he was drunk and there might have been trouble.

We spent the night at her mother's house and with the next morning the village people were ready to hit the fields and we were ready to hit the road. But before we could leave Nguyen's old grandfather wanted to talk to us and so her mom came along. Her grandfather had Nguyen and me hold hands and he said a bunch of words that I could not understand. After that, Nguyen told me we were married and that I was now part of their village. I had to grin (no paper signing). It was a hell of a lot easier for me than the first time around.

My original plans were to go back to Hue, spend a couple of days, put Nguyen on a train back to Nha Trang and I would be getting on a plane to Hanoi. Now that we were married and I was expected to consummate the marriage, my plans were changed and Nguyen would now be going to Hanoi with me. We spent a couple of great nights in Hue and then we flew to Hanoi. This was, of course, Nguyen's first plane trip and she was quite excited; so much so that she celebrated by throwing up during the flight. We spent a week in Hanoi with day long side trips to both Hoi Lu and Ha Long Bay. It was a great honeymoon..a honeymoon that I never expected. After the week was up I had to fly back to the states so I gave Nguyen some money and put her on a train back to Nha Trang.

Note: At this point Ron decided to share some intimate details with me, which he stated were for my eyes only. The comments were not sordid details, but about the kindness and tenderness that men and women can share… despite age and culture. Love is a universal song where the melody is understood by all.

Letter: April 29, 1999

Darling,

I had received your mail. That make me very happy. I like it very much, thank you about that.

Darling, my trip home on train is okey, but it very lonely without you. Now I come back to Nha Trang I went to work at same place and try to find another good work. I hope everythings are okey with you. I'm fine and my mother is too. It was so nice to be with my mother again, it was very nice and I thank you for that. The men in my village are sorry about my cousin and they wish to see you again. My mother tell me that every people in my village talk about you, nice I like it.

How about your father and your daughters? I hope they are well. Darling, I miss you, I love you and I hope you love me too.

All my love for you,

Nguyen

Letter May 27, 1999

Darling,

Today I'm so happy, happy that I can tell any words. I get your money and candy. That is my candy. I like it very much. I thank you very much and one thing was very big surprise that was my mother coming to Nha Trang to visit me because I was sick I had to go to hospital one week so my mother came to Nha Trang to look after me. But now I feel better so you don't worry darling.

Your friend Karl come to Nha Trang and stay here only 2 or 3 days so I don't have enough time to do something for you, I mean take the picture with my Ao Dai so I'll send to you later Darling, I write to you later and I write more and more but for now I put down my pen and I hope you stay in best of health. I think of you every day and miss you too.

All my Love for you,

Nguyen

Letter: August 23, 1999

My Darling,

It's 7 p.m. right now. Normally in the night I miss you, I don't know why so I write to you now. Darling, I thank you about your kindness you were worry about me that means you think of me every day and I'm very moved. I wish we can stay together so you can look after me and I can look after you.

Darling, I had stopped work in restaurant so in the evening I can go to learn something, I like it. I don't want stay at home it makes me feel very lonely and I miss you more. Before I worked a lot so I didn't have time to write to you but now I have time so I can converse with you but I don't know how to tell you how much I miss you. I miss your kiss, your beautiful eyes and kind smile. When I miss you I feel more lonely.

Darling, I was very happy that my mother had come to visit and look after me when I was in the hospital so I don't need go to Hue to visit her. When we were together we talk about you a lot and I talk about our future to my mother and my mother said she looking forward that day, she is happy too. My mother said "do you want to live in my village" and do same things the people in my village do? I said yes in the future when Ronald come to Vietnam I'll tell him about that and maybe we can live over here. My mother said: If Ronald do like farmer only one day he'll die and tomorrow no more Ronald. We talk alot of fun about our life, you know? I said to her that you don't have children anymore. First she was sad but then I explain to her some things and finally she understands and she said: Okey in the future you and Ronald have your own house and have two dogs and five cats. We had a nice time together darling. I go to bed now. goodbye my darling. I'm missing you.

Nguyen

Ron's Commentary:

My last letter from Nguyen

October. 10, 1999

Darling,

I've just gotten your mailing. I'm so happy that I write to you at once. You know I like the dress very much. It is my size and the color is very nice too. This dress is my dreams because in Vietnam it is very expensive. You are so wonderful to me. Thank you very much...

Darling, you know that because I am only child I have to stay in Vietnam to care for my mother in her old age. I just talk to my mother about our love and our future and my mother gave her ideas. Some of them is just joking so you don't worry so much because we have time to think of it and we have to earn some more money for our life in the future. I'm sorry because I can't earn more money for our life, but I will still try and try. I can tell you the price for building a nice home in the village is cheeper than Nha Trang. With $5,000

U.S.D. we can build a good home, but it would be $15,000 U.S.D. in Nha Trang. My mother want for us to live in village and I think you understand the Vietnam customs and that is why, but we can live everywhere you want, that is no problem, but if you live in the village you do not have to work because you are too old. If you want to have a small garden than that is okey.

Darling, you shouldn't buy many things for me because it cost alot of your money. You should save money for your life and take care of yourself. I'm okey, I have you that is very lucky for me it is enough for me. I'm worry about your tired because of me. I'm worry about you and think of you every day. I'm worry one day you forget me.

I always love you,

Nguyen

NOTE: The country of Vietnam was hammered when Typhoon Lingling struck in early November 1999 after visiting destruction through the Philippines. As it approached the central coast of Vietnam the storm picked up energy from the warm waters of the South China Sea. Hue was one of the hardest hit cities that were in the path of the storm. Little farming villages just west of Hue were certainly devastated.

When the storm blew out it left 749 known dead, over 49,000 houses destroyed and an estimated economic loss of over 4,000 billion Vietnam dong; certainly a staggering toll.

Ron's Commentary:

Jerry... not long after this letter I received a phone call from Nguyen and I don't really remember the date because I have put it out of my mind, but when she called she was crying, and I had a hard time trying to understand her when she told me that there was a big typhoon in Vietnam, [that many] parts of the country were flooded, and that she had received word that her mother's village had been swept away in the flood and some of the people were swept to the high grounds and had no food, and she did not know if her mother was still alive.

I told Nguyen that I would wire her enough money in the morning so that she could take supplies for the people of her village, rent a boat and go find her mother.... She was still sobbing when she hung up. That was the last time I had heard from Nguyen.

I later received a letter from Op, the vendor in Nha Trang, who told me that the small boat that Nguyen was in turned over in the fast water and she could not swim.

Best Regards,

Ron

My Personal Note:

The powers of the unspoken and unwritten words in this story are just as powerful if not much more as the words that survive. Every time I read the exchange, I too am swept away by rivers unseen to place unknown, and I pray for Nguyen's eternal soul and the soul of Vietnam in all its beauty and anguish.

Postscript 2013

It's now forty-three years since I returned home in 1970, forty years since the Paris Peace Accords and thirty-eight years since the North Vietnamese ignored the Paris Peace Accords and the government of the United States turned belly-up and abandoned the peoples of South Vietnam to a savage destruction by the North Vietnamese of all that was, once upon a time, the Republic of Vietnam.

There are tons of horror stories of rapes, murder, re-education camps and other unspeakable tragedies as the people of the south abandoned their everything in Vietnam to head out to sea into a storm of the unknown.

Vietnam today is well populated and fed, the city of Ho Chi Minh, which is still called Saigon by most, is booming with high rise buildings, terrible traffic, high-end shopping and the return of the drug and prostitution professions.

But, as terrible as it may have been and in some ways it may still be, there is a zest for life in the south and a real like for Americans when we show up. The Europeans, mainly French and Germans, hunt and haunt the big cities and the beaches looking for the "vacation experience" they can talk about when they get home to knockwurst and cheese.

But, for the most part, the returning Americans are old men, like me, looking for the spots where they lived and their friends died as young men, trying to cope with the passage of time and to get their head straight before the final sleep.

I recently wrote a dear friend an email addressing this book I was writing, a book that is small but a book that took me ten years of inner struggles to bring to a conclusion.

What I told him was, in part:

> What I wrote is filled with lots of sex, drinking, parties and the haunting call of something unclear.... I did not intend for it to turn out that way, but it does, and after I read it in the entirety, I'm surprised at

what I wrote as a body of work... and I even deliberately left out lots of small pieces that I thought "too much"...

I never was and don't feel I am the saint you see... I am just a lucky, ordinary guy struggling through life with that still "something unclear" that follows me every day and lets me know that it is there, either as a reward or a punishment, I know not... something unclear.

Vietnam was a terrible war, a horrible time for many. It was also the source of a deep undeniable knowledge that, in reality, it was more than a war. For many others, though, Vietnam continues to be a land without a dream.

Jerry Harlowe
Writing from Catonsville, MD
2013